U. S. Navy Seabee Combat Handbook, Volume 2

Department of Defense

Pentagon
Publishing

NONRESIDENT
TRAINING
COURSE

March 1996

Seabee Combat Handbook, Volume 2

NAVEDTRA 14235

Although the words "he," "him," and "his" are used sparingly in this course to enhance communication, they are not intended to be gender driven or to affront or discriminate against anyone.

COMMANDING OFFICER
NETPTDC
6490 SAUFLEY FIELD ROAD
PENSACOLA, FL 32509-5237

Errata #2 07 May 01

Specific Instruction and Errata for
Nonresident Training Course

SEABEE COMBAT HANDBOOK, VOLUME 2, NAVEDTRA 14235

1. This errata supersedes all previous errata. No attempt has been made to issue corrections for errors in typing, punctuation, etc., that do not affect your ability to answer the question or questions.

2. To receive credit for deleted questions, show this errata to your local course administrator (ESO/scorer). The local course administrator is directed to correct the course and the answer key by indicating the questions deleted.

3. Assignment Booklet, NAVEDTRA 14235

Delete the following questions and leave the corresponding spaces blank on the answer sheets:

Question

1-43

2-10

2-56

3-34

4. Change the following items in the Assignment Booklet, NAVEDTRA 14235:

a. Question 4-9 page 19; change the question's stem to read "The depth of a platoon defense is the distance between the squad's primary positions and the rearward extension of what other position?"

b. Question 4-57 page 23; change the question's stem to read "... occupied by the convoy commander?"

C. Change figure 5B, item C, page 27 to read "Immediate Assault Drill."

d. Question 6-33 page 33; change the question's stem to read "... conducting detailed troop decon (DTD)?"

e. Question 6-46 page 34; change the question's stem to read "... a sanitizing solution at station 7, ..."

f. Question 6-58 page 35; change the question's stem to read "... the primary task of DED station 4?"

PREFACE

By enrolling in this self-study course, you have demonstrated a desire to improve yourself and the Navy. Remember, however, this self-study course is only one part of the total Navy training program. Practical experience, schools, selected reading, and your desire to succeed are also necessary to successfully round out a fully meaningful training program.

THE COURSE: This self-study course is organized into subject matter areas, each containing learning objectives to help you determine what you should learn along with text and illustrations to help you understand the information. The subject matter reflects day-to-day requirements and experiences of personnel in the rating or skill area. It also reflects guidance provided by Enlisted Community Managers (ECMs) and other senior personnel, technical references, instructions, etc., and either the occupational or naval standards, which are listed in the *Manual of Navy Enlisted Manpower Personnel Classifications and Occupational Standards*, NAVPERS 18068.

THE QUESTIONS: The questions that appear in this course are designed to help you understand the material in the text.

VALUE: In completing this course, you will improve your military and professional knowledge. Importantly, it can also help you study for the Navy-wide advancement in rate examination. If you are studying and discover a reference in the text to another publication for further information, look it up.

1996 Edition Prepared by
EOC Ricky P. Baptista

Published by
NAVAL EDUCATION AND TRAINING
PROFESSIONAL DEVELOPMENT
AND TECHNOLOGY CENTER

NAVSUP Logistics Tracking Number
0504-LP-026-8730

Sailor's Creed

"I am a United States Sailor.

I will support and defend the Constitution of the United States of America and I will obey the orders of those appointed over me.

I represent the fighting spirit of the Navy and those who have gone before me to defend freedom and democracy around the world.

I proudly serve my country's Navy combat team with honor, courage and commitment.

I am committed to excellence and the fair treatment of all."

CONTENTS

SUMMARY OF SEABEE COMBAT HANDBOOK TRAINING MANUALS

VOLUME 1

Seabee Combat Handbook, Volume 1, NAVEDTRA 12003, consists of chapters on the History and Organization of the Seabees and Laws of War; Special Clothing and Equipment; Service Rifle and Pistol Marksmanship; Combat Maneuvers, Formation, Patrols, and Ambushes; Land Navigation; Evasion, Survival, and Escape; Individual Protection Measures; Entanglements; Chemical, Biological, and Radiological (CBR) Defense; First Aid and Field Sanitation; Organic Communications Equipment; Hand Grenades, Land Mines, and Booby Traps; Organic Support Weapons: M203 and Machine Guns; and Organic Support Weapons: 60mm Mortar and AT4.

VOLUME 2

Seabee Combat Handbook, Volume 2, NAVEDTRA 12004, consists of chapters on Organization and Operation of the Combat Operations Center; Organization and Operation of the Company Command Post; Setup and Control of Medical Evacuation (MEDEVAC); Planning and Development of Defense Tactics; Counter Ambush Techniques; and CBR Decontamination.

INSTRUCTIONS FOR TAKING THE COURSE

ASSIGNMENTS

The text pages that you are to study are listed at the beginning of each assignment. Study these pages carefully before attempting to answer the questions. Pay close attention to tables and illustrations and read the learning objectives. The learning objectives state what you should be able to do after studying the material. Answering the questions correctly helps you accomplish the objectives.

SELECTING YOUR ANSWERS

Read each question carefully, then select the BEST answer. You may refer freely to the text. The answers must be the result of your own work and decisions. You are prohibited from referring to or copying the answers of others and from giving answers to anyone else taking the course.

SUBMITTING YOUR ASSIGNMENTS

To have your assignments graded, you must be enrolled in the course with the Nonresident Training Course Administration Branch at the Naval Education and Training Professional Development and Technology Center (NETPDTC). Following enrollment, there are two ways of having your assignments graded: (1) use the Internet to submit your assignments as you complete them, or (2) send all the assignments at one time by mail to NETPDTC.

Grading on the Internet: Advantages to Internet grading are:

- you may submit your answers as soon as you complete an assignment, and
- you get your results faster; usually by the next working day (approximately 24 hours).

In addition to receiving grade results for each assignment, you will receive course completion confirmation once you have completed all the assignments. To submit your assignment answers via the Internet, go to:

https://courses.cnet.navy.mil

Grading by Mail: When you submit answer sheets by mail, send all of your assignments at one time. Do NOT submit individual answer sheets for grading. Mail all of your assignments in an envelope, which you either provide yourself or obtain from your nearest Educational Services Officer (ESO). Submit answer sheets to:

COMMANDING OFFICER
NETPDTC N331
6490 SAUFLEY FIELD ROAD
PENSACOLA FL 32559-5000

Answer Sheets: All courses include one "scannable" answer sheet for each assignment. These answer sheets are preprinted with your SSN, name, assignment number, and course number. Explanations for completing the answer sheets are on the answer sheet.

Do not use answer sheet reproductions: Use only the original answer sheets that we provide—reproductions will not work with our scanning equipment and cannot be processed.

Follow the instructions for marking your answers on the answer sheet. Be sure that blocks 1, 2, and 3 are filled in correctly. This information is necessary for your course to be properly processed and for you to receive credit for your work.

COMPLETION TIME

Courses must be completed within 12 months from the date of enrollment. This includes time required to resubmit failed assignments.

PASS/FAIL ASSIGNMENT PROCEDURES

If your overall course score is 3.2 or higher, you will pass the course and will not be required to resubmit assignments. Once your assignments have been graded you will receive course completion confirmation.

If you receive less than a 3.2 on any assignment and your overall course score is below 3.2, you will be given the opportunity to resubmit failed assignments. **You may resubmit failed assignments only once.** Internet students will receive notification when they have failed an assignment--they may then resubmit failed assignments on the web site. Internet students may view and print results for failed assignments from the web site. Students who submit by mail will receive a failing result letter and a new answer sheet for resubmission of each failed assignment.

COMPLETION CONFIRMATION

After successfully completing this course, you will receive a letter of completion.

ERRATA

Errata are used to correct minor errors or delete obsolete information in a course. Errata may also be used to provide instructions to the student. If a course has an errata, it will be included as the first page(s) after the front cover. Errata for all courses can be accessed and viewed/downloaded at:

https://www.advancement.cnet.navy.mil

STUDENT FEEDBACK QUESTIONS

We value your suggestions, questions, and criticisms on our courses. If you would like to communicate with us regarding this course, we encourage you, if possible, to use e-mail. If you write or fax, please use a copy of the Student Comment form that follows this page.

For subject matter questions:

E-mail: n314.products@cnet.navy.mil
Phone: Comm: (850) 452-1001, Ext. 1826
DSN: 922-1001, Ext. 1826
FAX: (850) 452-1370
(Do not fax answer sheets.)
Address: COMMANDING OFFICER
NETPDTC (CODE N314)
6490 SAUFLEY FIELD ROAD
PENSACOLA FL 32509-5237

For enrollment, shipping, grading, or completion letter questions

E-mail: fleetservices@cnet.navy.mil
Phone: Toll Free: 877-264-8583
Comm: (850) 452-1511/1181/1859
DSN: 922-1511/1181/1859
FAX: (850) 452-1370
(Do not fax answer sheets.)
Address: COMMANDING OFFICER
NETPDTC (CODE N331)
6490 SAUFLEY FIELD ROAD
PENSACOLA FL 32559-5000

NAVAL RESERVE RETIREMENT CREDIT

If you are a member of the Naval Reserve, you will receive retirement points if you are authorized to receive them under current directives governing retirement of Naval Reserve personnel. For Naval Reserve retirement, this course is evaluated at 9 points. (Refer to *Administrative Procedures for Naval Reservists on Inactive Duty,* BUPERSINST 1001.39, for more information about retirement points.)

COURSE OBJECTIVES

In completing this nonresident training course, you will demonstrate a knowledge of the subject matter by correctly answering questions on the following: Organization and Operation of the Combat Operations Center; Organization and Operation of the Company Command Post; Setup and Control of Medical Evacuation (MEDEVAC); Planning and Development of Defense Tactics; Counter Ambush Techniques; and CBR Decontamination.

Student Comments

Course Title: *Seabee Combat Handbook, Volume 2*

NAVEDTRA: 14235 **Date:** ____________________

We need some information about you:

Rate/Rank and Name: ____________________ SSN: ____________ Command/Unit ________________

Street Address: ________________________ City: ____________ State/FPO: ________ Zip _______

Your comments, suggestions, etc.:

Privacy Act Statement: Under authority of Title 5, USC 301, information regarding your military status is requested in processing your comments and in preparing a reply. This information will not be divulged without written authorization to anyone other than those within DOD for official use in determining performance.

NETPDTC 1550/41 (Rev 4-00)

CHAPTER 1

ORGANIZATION AND OPERATION OF THE COMBAT OPERATIONS CENTER

Efficiently organized and executed command and control operations are essential for any combat operation to be successful. Many battles have been lost because of poor execution of command and control activities. A major reason for the success of Operation Desert Storm was the manner each unit maintained and supported a vigorous command and control structure.

A comprehensive knowledge about higher and lower echelons in the area of operations assigned to a Seabee battalion is crucial to personnel executing command and control functions. This chapter provides detailed information on the Combat Operations Center (COC). Also, described in this chapter is the Marine Air-Ground Task Force (MAGTF).

NOTE: During combat operations, Seabee units are normally assigned to a MAGTF.

ORGANIZATION OF THE MARINE AIR-GROUND TASK FORCE (MAGTF)

A Marine Air-Ground Task Force (MAGTF) is normally formed for combat operations in which substantial Marine aviation and Marine ground units participate. Trends in national strategy show that the most probable employment of MAGTFs will be in the execution of force-in-readiness missions. Accomplishment of such missions requires the employment of MAGTFs as landing forces in amphibious operations and later operations ashore. The MAGTFs must have the capability in such commitments to operate in close coordination with other U.S. Armed Forces and the forces of Allied nations. MAGTFs are characterized by operational concepts, organizational structures, equipment, and systems that are suited for employment in Navy-Marine, joint, or combined operations.

The fact that a Seabee battalion is attached to a MAGTF during combat makes it critical that you understand the MAGTF organization. No matter how large, a MAGTF consists of at least four elements with seven primary engineer (fig. 1-1) assets spread out among the elements.

The four elements (fig. 1-2) are as follows:

- Command Element (CE)
- Ground Combat Element (GCE)
- Aviation Combat Element (ACE)
- Combat Service Support Element (CSSE)

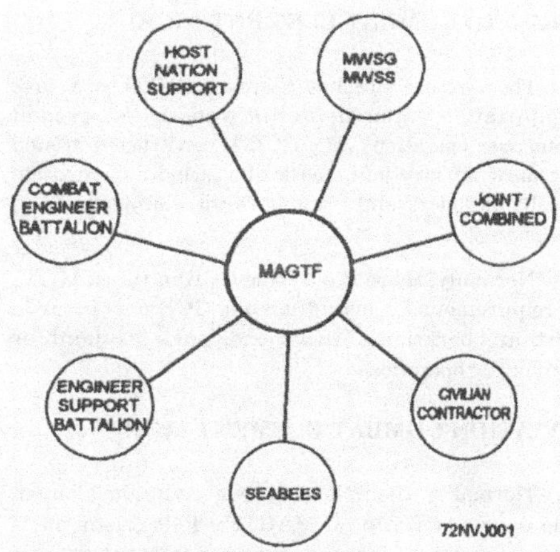

Figure 1-1.—Primary MAGTF Engineer Assets.

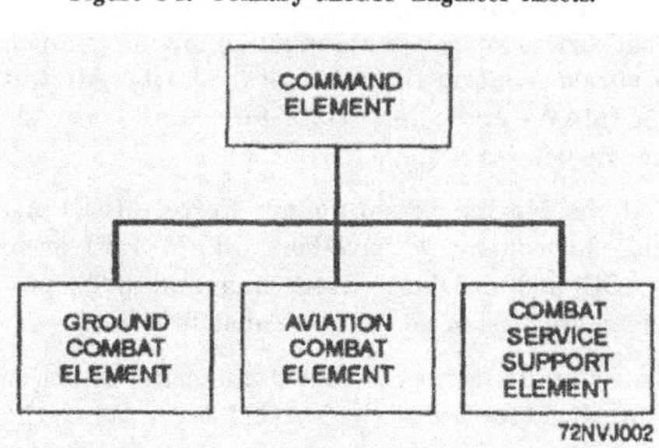

Figure 1-2.—Elements of a MAGTF organization.

COMMAND ELEMENT (CE)

The Command Element (CE) is the MAGTF headquarters and is established for effective planning and execution. It extends and complements the capabilities of subordinate MAGTF elements but do not duplicate them under normal circumstances. Direct liaison among the four elements is desirable to achieve the necessary coordination required for the effective conduct of air-ground operations.

A CEC officer is attached to the Command Element, engineer staff. The CEC officer serves as an advisor to the MAGTF engineer officer on matters relating to the capabilities and employment of the Naval Construction Force. A large portion of the MAGTF Command Element is concerned with matters involving higher, adjacent, and supporting commands.

GROUND COMBAT ELEMENT (GCE)

The Ground Combat Element (GCE) is a task organization tailored for the conduct of ground maneuver operations. The GCE is constructed around a combat infantry unit, and it also includes appropriate combat support and combat service support (CSS) units.

Normally, there is only one GCE in the MAGTF. A requirement for more than one GCE may occur in certain operational situations, such as joint or combined operations.

AVIATION COMBAT ELEMENT (ACE)

Normally, there is only one Aviation Combat Element (ACE) in a MAGTF. This element is task-organized for the conduct of tactical air operations. It includes the aviation commands (including air control agencies), combat support, and combat service support units required for the mission. The varied aviation resources of a Marine Aircraft Wing (MAW) and appropriate force units provide these capabilities.

At the Marine Expeditionary Force (MEF) and Marine Expeditionary Force-Forward (MEF-F) level, the ACE includes both fixed-wing and helicopter assets as well as an air defense capability.

Air operations are conducted under the principle of centralized control at the MAGTF level. When the MAGTF commander assumes responsibility for control of air operations, he or she exercises control through facilities provided by the ACE. He normally appoints the commander of the ACE to act as the MAGTF tactical air commander (TAC).

COMBAT SERVICE SUPPORT ELEMENT (CSSE)

The Combat Service Support Element (CSSE) provides the combat service support (CSS) to the MAGTF that is beyond the organic capability of the subordinate elements. Depending on the assigned mission, it is task organized to provide all of the following functions: supply, maintenance, engineer, medical/dental, automated data processing, material-handling equipment, personal services, food services, transportation, military police, disbursing, and financial management.

SEPARATE TASK ORGANIZATIONS

Although there are always four major elements within a MAGTF, separate task organizations required to perform combat support or combat service support functions may be formed. Naval Construction Force Units come under this type of organization. Under these circumstance, the NCF is OPCON (see Glossary) to the MAGTF commander within the Command Element (fig. 1-3).

TYPES OF MAGTFs

The type of MAGTF Seabees would be supporting is dictated by the contingency. The three types of MAGTFs are as follows:

- Marine Expeditionary Unit (MEU)
- Marine Expeditionary Force-Forward (MEF-F)
- Marine Expeditionary Force (MEF)

MARINE EXPEDITIONARY UNIT (MEU)

The Marine Expeditionary Unit (MEU) is a task organization that is normally commanded by a colonel and is capable of performing combat operations of a limited scope. The MEU is the air-ground team organization that is normally used for routine, forward afloat, and deployment requirements. The MEU provides an immediate reaction capability to crises and, when committed, is normally supported from its sea base. Normally, an Air Det from a Seabee battalion is assigned to an MEU (fig. 1-4).

The MAGTFs elements for a MEU are as follows:

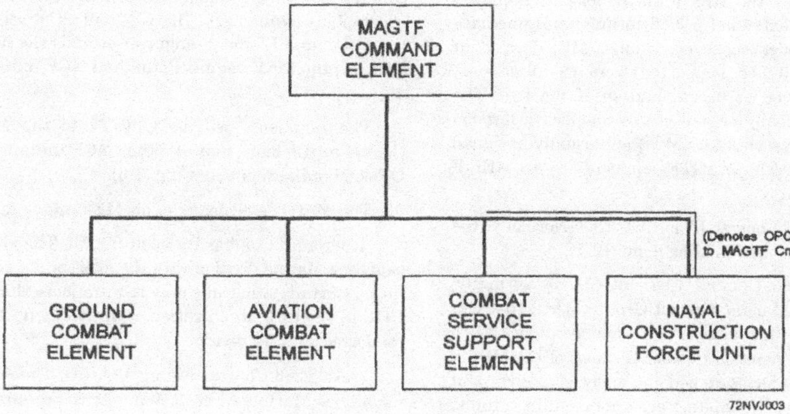

Figure 1-3.—MAGTF command relationships.

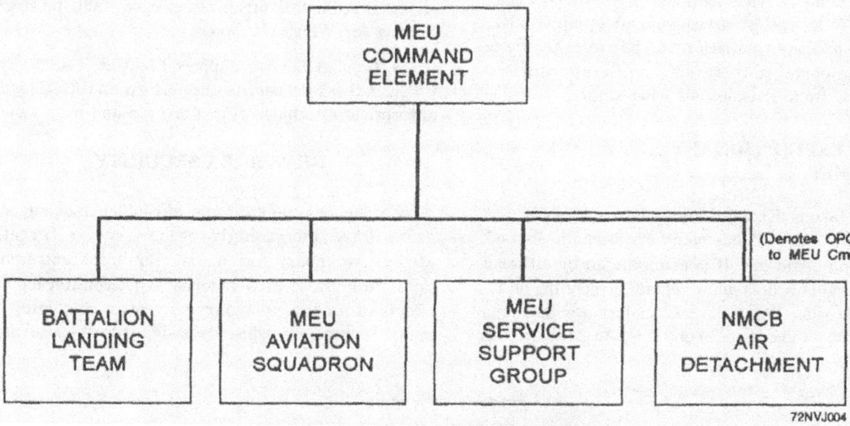

Figure 1-4.—MEU command relationships.

1. Ground Combat Element (GCE): Normally, a battalion landing team (BLT). Only under unusual cinnunstances would the GCE consist of two BLTs.

2. Aviation Combat Element (ACE): Normally, a composite helicopter squadron. However, in certain instances, the ACE may consist of an attack squadron, a helicopter squadron, and elements of an observation squadron.

3. Combat Service Support Element (CSSE): Within an MEU, normally called a service support group (MSSG). The MSSG is task-organized primarily from the force service support group (FSSG) with appropriate attachments from division and wing assets.

MARINE EXPEDITIONARY FORCE-FORWARD (MEF-F)

The Marine Expeditionary Force-Forward (MEF-F) (formerly called Marine Expeditionary Brigade [MEB]) is a task organization that is normally commanded by a brigadier general. An MEF-F is capable of conducting operations in low- and mid-intensity conflict environments. During potential crises, the MEF-F maybe forward deployed afloat for an extended period to provide immediate response. Under these conditions, MEF-F combat operations may be supported from the sea base, facilities ashore, or a combination of the two. The MEF-F is normally organized to wcomplish a mission of limited scope and an NMCB is normally assigned to an MEF-F. The MAGTFs elements of an MEF-F are as follows:

1. Ground Combat Element (GCE): Normally, the GCE is a regimental landing team (RLT).

2. Aviation Combat Element (ACE) Normally, the ACE is a Marine Aircraft Group (MAG) that can provide all of the functions of Marine aviation. Unlike the MEU, the Aviation Combat Element of the MEF-F is organized and equipped for early establishment ashore. Should the landing area not contain suitable airfields, an expeditionary airfield will be developed using assets organic to the MEF-F.

3. Combat Service Support Element (CSSE): The CSSE is a brigade service support group (BSSG). The BSSG is task-organized primarily from the force service support group (FSSG) with appropriate attachments from division and wing assets.

MARINE EXPEDITIONARY FORCE (MEF)

The Marine Expeditionary Force (MEF), the largest of the MAGTF, has many variations in its task organization structure. It is commanded by either a major general or a lieutenant general, depending on its size and mission. The MEF can conduct a wide range of amphibious and ashore operations. An MEF can be tailored for any intensity of combat and to any geographic environment. The CEC officer attached to the Command Element, engineer staff, is the liaison between the MEF engineer staff and NCF units (fig. 1-5).

The regiments will be OPCON to the MEF's Command Element and all other NCF units will be OPCON to the regiments (fig. 1-6).

The MAGTFs elements of an MEF are as follows:

1. Ground Combat Element (GCE): The GCE is usually a Marine division with the appropriate combat units. Certain situations may require more than one GCE in the task organization, conceivably up to two reinforced Marine divisions.

2. Aviation Combat Element (ACE): The ACE is usually a Marine Aircraft Wing task organized to conduct all types of tactical air operations. The element is designed for operations in an expeditionary environment. Certain situations may require more than one ACE in the task organization, conceivably up to two Marine Air Wings (MAWs).

3. Combat Service Support Element (CSSE): The CSSE is the force service support group (FSSG) with appropriate attachments from division and wing assets.

REAR AREA SECURITY

Throughout the spectrum of conflict, rear support areas have increasingly become major targets. Modern weaponry has made rear areas extremely vulnerable. Rear area security must counteract the effects of modem weaponry to protect rear support areas. Seabees are primarily assigned to the rear area

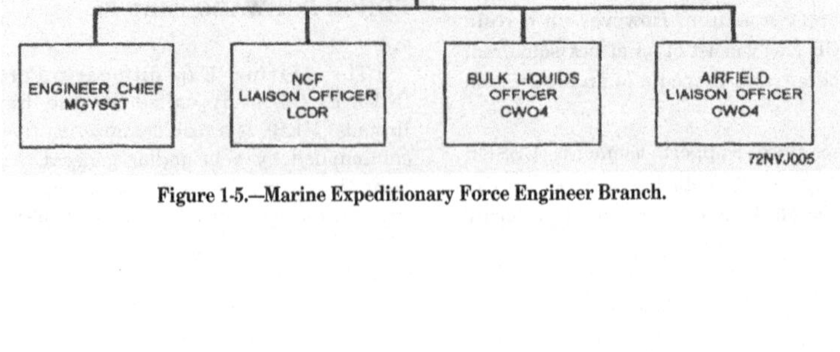

Figure 1-5.—Marine Expeditionary Force Engineer Branch.

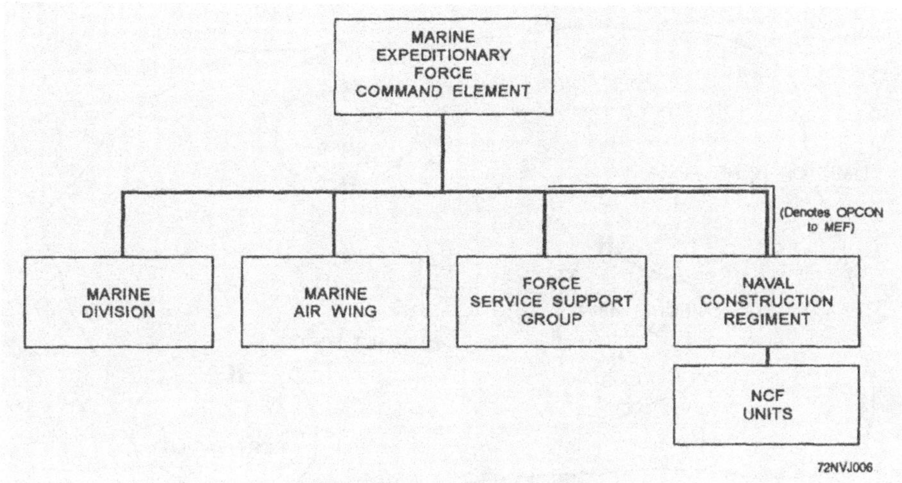

Figure 1-6.—MEF major subordinate commands.

and therefore must be familiar with the rear area organization.

The main objective of rear area security (RAS) is to minimize the effects of an enemy attack. Tactics used by the RAS include measures taken prior, during, and after an enemy airborne attack, sabotage, infiltration, guerrilla action, or initiation of psychological or propaganda warfare. The MAGTF rear area is the area extended rearward from the rear boundary of the Ground Combat Element (GCE) to the MAGTF rear boundary.

The objective of RAS is to provide defense for all forces operating within the MAGTF rear area so those functions associated with rear area operations, in support of combat operations, are not interrupted. All units in the rear area must be prepared to defend themselves to accomplish this objective. A cluster concept is used within the rear area so base facility operations centers are established. An NMCB has a formidable array of weapons and has a significant capability for defensive operations within the rear area.

BASE FACILITY COMMANDER

The base facility commander is responsible for the facility assigned to him or her. Each unit is assigned to a base facility and reports to the base facility commander by way of the COC (fig. 1-7).

REAR AREA SECURITY COORDINATOR (RASC)

The CSSE or the ACE commander is usually appointed by the MAGTF commander to coordinate rear mea security (RAS) and normally functions as the rear area security coordinator (RASC). The RASC monitors the day-today operations of the rear area through the combat service support operations center (CSSOC) and the rear area operations center (RAOC). The NCF commander coordinates with the RASC to ensure that NCF assets and capabilities are incorporated into the RAS effort.

PROVISIONAL SECURITY FORCES

An MEF RASC can organize two types of provisional security forces: the previsional mobile security platoons (PMSPs) and the provisional helicopter-borne security company (PHSC). MAGTFs, smaller than an MEF, will have a tailored provisional security force. Based on the mission and threat assessment, these units can be used to support local defense efforts in the support of the RAS. These units should be included in your defense plan.

Provisional Mobile Security Platoons (PMSPs)

The force service support group (FSSG), located within the rear area, can field two provisional mobile

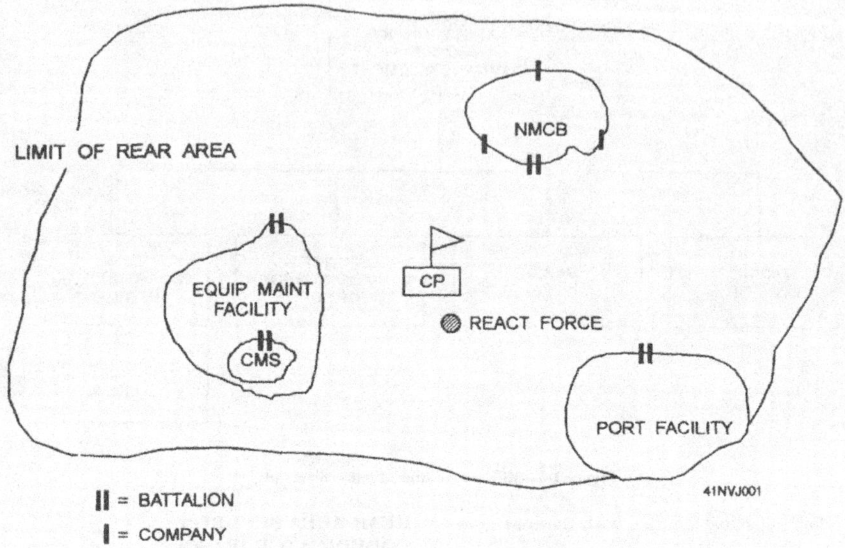

Figure 1-7.—Security of base facilities.

security platoons (PMSPs) to act as a quick reaction force in support of RAS efforts. The PMSPs can be tasked with the following missions:

- Relief/rescue of attacked installations/units
- Route patrolling and convoy protection
- Surveillance/reconnaissance
- Defense of possible enemy drop/landing zones
- Finding, fixing, destroying enemy forces operating in the rear area

Force Service Support Group Military Police

The force service support group military police are the MPs for the rear area and can be tasked with the following missions:

1. Provide military police to conduct battlefield circulation control for the MAGTF (material supply routes security).

2. Provide military police for law enforcement, criminal investigation, U.S. prisoner confinement, and counteract terrorist activities.

3. Establish surveillance and conduct route reconnaissance in the MAGTF rear area.

4. Provide for the collection, processing, and evacuation of enemy prisoners of war (POW) and civilian internees in the MAGTF rear am.

COMBAT OPERATIONS CENTER (COC)

The COC is established to provide the battalion commander centralized command and control facilities for all combat tactical and operational operations conducted under his or her command. It is the focal point for the battalion and the terminating point for all tactical and nontactical radio nets.

The primary purpose of the COC is to monitor and record the tactical and nontactical operations of the battalion continually. The specific composition and functions of the COC will vary with the desires of the individual commander. Normal COC functions include the following:

- Receiving and recording operational reports from subordinate elements and companies
- Maintaining current plots of the friendly and enemy situations and displaying this information within the COC

- Preparing and submitting operational reports to higher headquarters
- Providing dedicated communication channels for tactical and operational reporting
- Transmitting orders and tactical decisions of the battalion commander to companies subordinate elements, and higher headquarter as required
- Monitoring the progress of the battalion's tactical operations and expeditiously reporting significant events or incidents to the S-3 or the commander
- Advising the fire support coordination center (FSCC) and interested staff sections of events or infomation of immediate concern to them
- Serving as the principal point of contact for liaison personnel from subordinate, supporting, or adjacent tactical elements

The COC controls the battalion's tactical nets established by higher headquarters. A COC normally has direct sole-user telephone circuits (hot lines) to major subordinate tactical units and to the COC of higher headquarters. Besides voice radio nets and telephones, the battalion's COC normally maintains direct teletype links with major subordinate elements and with the COC of higher headquarters.

The operations officer (S-3) exercises staff supervision over the COC in coordination with the training (S-7) and intelligence officer (S-2).

LOCATION OF THE COMBAT OPERATIONS CENTER (COC)

When establishing a COC, you should give special consideration to location. The location of the COC must be hard to detect by enemy forces, easy to defend, and be established in a centralized location that is easily accessible to battalion personnel. Other factors to consider are centralized communications and vehicle traffic.

Centralized Communications

The main COC is centrally located to each company's command post to ensure positive control by the battalion commander.

Vehicle Traffic

As previously stated, the COC is the principal point of contact for liaison personnel from subordinate, supporting, or adjacent tactical elements. Strict traffic control measures are essential when these supporting elements visit the COC. These include the following:

- At least two dismount and two entry points are used to reduce traffic concentrations.
- Passenger dismount points are concealed from direct observation.
- Vehicles are parked and camouflaged in one or more parking areas within the local security perimeter of the COC.
- Construction of new roads into or by the COC is normally prohibited. To lessen the chance of detection by the enemy, you can use existing roads.
- Vehicles entering or departing the area of the COC are required to use exits that are concealed by are cover or camouflage.

COMMUNICATIONS

Superior communication is critical to COC operations. The location of the COC is one of the factors that determines the communication assets required when establishing a communication plan. Also affected by the location of the COC is the location of the "Antenna Farm" discussed later in this chapter.

DEFENSE AND SECURITY

To protect the COC from a direct attack or enemy infiltration, you must consider vital defense and security measures, such as the following:

1. Security force: A well-trained and organized security force assigned to the COC.

2. Terrain that enhances security: An area of irregular, well-forested ground hampers aerial observation and, if it includes high brush and low trees as well, it makes enemy ground observation more difficult. Full use is made of natural cover, concealment, and supplementary camouflage measures. Individual camouflage discipline is rigidly enforced.

3. Location: Located near a unit that can assist in furnishing security.

4. Entanglements: Barbed and tactical wire entanglements.

5. Night noise and light discipline: Ensure the generator supplying power to the COC is located a good distance from the COC and is sandbagged to muffle the

sound of the generator. Also, ensure the tent used for the COC maintains light discipline.

ALTERNATE COC

In case the COC is destroyed, an alternate COC is established. All the consideration for a COC am duplicated in the alternate COC. Location of the alternate COC is away from the main COC and security is similar to the main COC. In the event the main COC is destroyed, operations are expediently assumed in the alternate COC.

ORGANIZATION OF THE COMBAT OPERATION CENTER

The internal organization of the COC may vary with each battalion commander or contingency operations. Written Standard Operating Procedures (SOPS) are established to avoid any confusion (fig. 1-8). Personnel duties, maps, and various boards are usually standard within any COC organization.

PERSONNEL AND THEIR DUTIES

The operations officer (S-3) is primarily responsible for coordinating, organizing, operating,

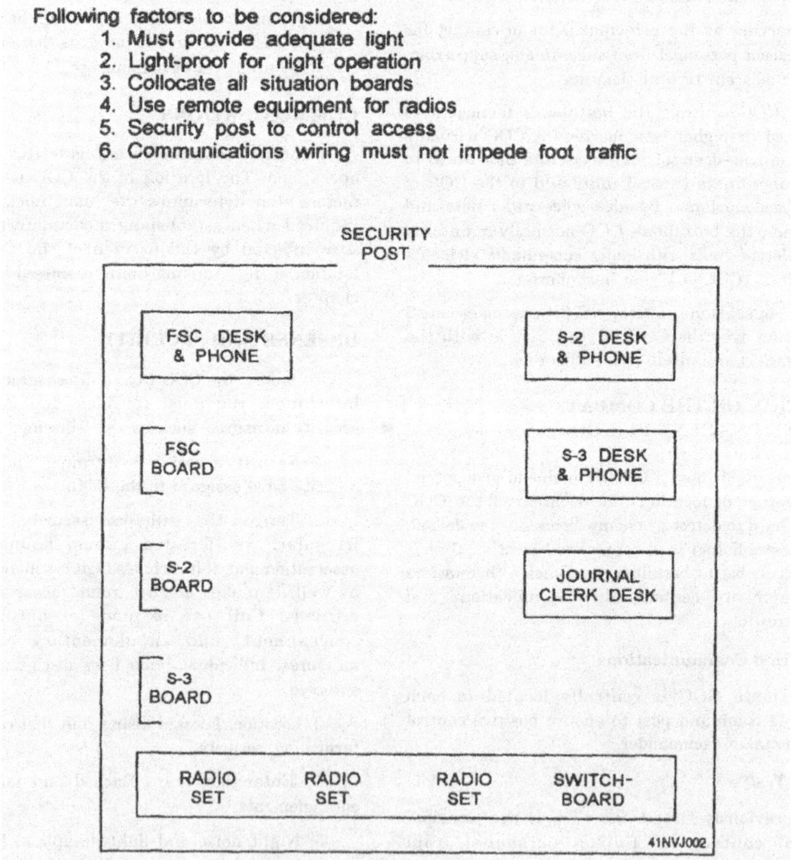

Figure 1-8.—Internal arrangement of COC.

- Assists the section watch officer as required.

NBC/CBR Officer

The NBC/CBR officer plays a critical role in the early warning of a nuclear, biological, or chemical attack. The NBC/CBR officer should hold the proper Naval Enlisted Classification Code (NEC) to fill this position proficiently. The NBC/CBR officer's duties include the following:

- Responsible for organizing, training, and supervising personnel assigned to the battalion's decontamination teams.
- Monitors all incoming messages related to weather information.
- Responsible for all outgoing and incoming NBC reports.
- Establishes NBC fallout zones to provide early warning of an NBC attack.
- Advises the section watch officer on the setting of mission-oriented protective posture levels (MOPP levels).

CHARTS AND MAPS

The types of charts and maps required within the COC vary according to the mission and the battalion commander. All charts and maps must be visible to all personnel in the COC. Ensuring proper placement enables all personnel standing watch in the COC to monitor current situations and to respond to any incoming messages expediently. The maps and charts provide detailed operations and INTEL information. Also posted in the COC is a clipboard for an incoming/outgoing message reading file.

The operation map shows the area of operation and should include the following:

- Friendly troop positions including the battalion's fire plan
- Current locations of command posts in the area of operation
- Location and status of patrols
- Landing zone locations
- NBC corridors/fallouts

The operation charts should show the following:

- Personnel strength
- POWs or EPWs
- Report status
- Equipment status
- Call signs and requencies
- Convoy status
- Casualties wounded/KIA
- Food/water/ammo status

The INTEL map should show the following:

- Enemy troop disposition/strength within the AO
- Enemy equipment/weapons
- Weather and astronomical data
- NBC corridors/fallouts

The INTEL charts should show the following:

- Current INTEL
- Challenge and passwords
- Current MOPP level

COMMUNICATION PLAN

On a modern battlefield, the ability of a battalion commander to pass information between his or her troops and superiors is critical for the success of any mission. An uncomplicated, reliable, flexible, and responsive communication plan will enable the battalion commander to carry out his or her operational plans effectively. A well-thought-out communication plan is paramount to command and control. Communication is the instruments by which a battalion commander makes his or her orders known and, as such, is the voice of the command. Communication permits the battalion commander to exercise command and control of assigned forces, supporting fires, and combat service support over larger areas. Any transmission speaks only for and with the authority of the battalion commander who originates the transmission. A secondary purpose of communications is to simplify the transfer of information between individuals and groups of individuals that is necessary to the exercise of command and control.

Since the COC is the nerve center for the battalion and the terminating point for all tactical and nontactical radio nets, a well-thought-out communication plan must be established. Like everything else, the communication plan is based upon the mission. The communication officer is usually tasked with the development of the communication plan. He or she should include in the plan communication requirements for the COC, the setup of a remote antenna farm, and a communication network.

COMMUNICATION REQUIREMENTS

Communication requirements within the COC may vary according to the mission but the communication equipment used to fulfill the requirements is the same. Communication equipment will include one radio per operating net and field telephones (TA-312/PT) for S-3, S-2, and FSC. In general the requirements are as follows:

1. AN/PRC-77 or AN/PRC-119A: Radios used to communicate with the base facility commander, company command posts, convoys, forward observer, and other units located within the base facility.

2. TA-312/PT (Field Phone): A direct phone line to the company command posts and the Fire Direction Center (FDC battalion mortars).

3. AN/PRC-104: A high-frequency (GF) radio usually used to communicate with higher authority outside the base facility.

4. SB-22/PT: A switchboard used to connect numerous TA-312s.

5. AN/GRA-39: A remote unit that enables the operator to transmit and receive voice communication through a radio set from a distance of up to 2 miles. It is used to link radio communication from a remote antenna farm to the COC.

ANTENNA FARM

The antenna farm or communication site is the primary location of various antennas and communication equipment needed to establish communication in the field. Special consideration is given for selecting a location for the antenna farm because all radio communication in the COC depends on it.

An important consideration in selecting your site is accessibility. Time should not be wasted by establishing accessibility. Whenever possible, a site should be located near good reads. This will minimize any difficulty in supplying the site with water, fuel, oil, food, and ammunition.

Avoid obstructions like steel bridges, underpasses, power lines, or power units that can cause a weak or distorted signal coming from your communication equipment. Better results are obtained when the antennas are high and clear of hills, cliffs, buildings, densely wooded areas, and other obstructions.

Other factors that must be considered are physical security and a location where terrain will not interfere with transmissions. A relatively flat hilltop is usually the most desirable site location.

COMMUNICATION NETWORK

A communication network is a written detailed plan accompanied with a wire and radio plan detailing where each type of communication equipment transmits to (figs. 1-9 and 1-10).

To produce a communication network, you must know what frequency nets will be required for the mission. Higher authority will issue a Communication Electronic Operating Instruction (CEOI) to assist you with determining the frequency nets required.

CEOI contains the technical guidance required to establish and maintain communications in support of operations. The CEOI provides the details required to coordinate and control the various communications means and functions within a unit. This document normally contains call signs, call words, and frequencies to be used by designated operating units. By providing a standardized source of information, the CEOI enhances both operational communications capabilities and communications security. In case of loss or compromise, the standardized format simplifies rapid identification, destruction, and replacement. The following information and instructions are normally included in the CEOI:

- General communications instructions
- Call sign assignments
- Frequency assignments
- Radio net circuit designator
- Wire/cable trunk circuit designations
- Wire/cable tagging codes
- Telephone directory names and numbers
- Teletype and data muting indicators
- Identification and marking panel codes
- Signal panel message instructions
- Pyrotechnic and smoke codes
- Ground-air signals
- Sound warning signals

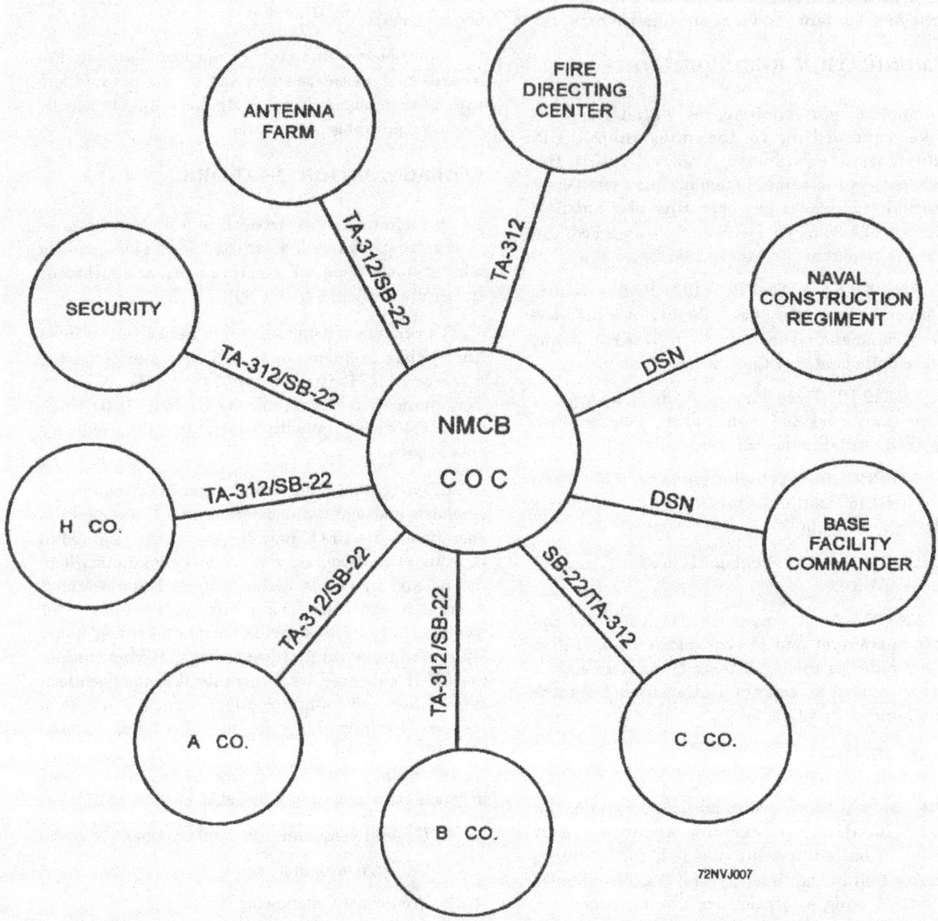

Figure 1-9.—COC wire plan.

- General cryptographic instructions
- Cryptographic devices to be used
- Current effective editions of the cryptographic key lists
- Codes and ciphers
- Passwords/challenge
- General authentication instructions
- Effective authentication tables

Each CEOI is classified according to content. General classification of the CEOI is based on the highest classified instruction it contains. Some of the frequency nets that may be listed in the CEOI and may be required in the COC areas follows:

1. Battalion Command Net: Establishes communication to the companies for administrative and logistics issues.

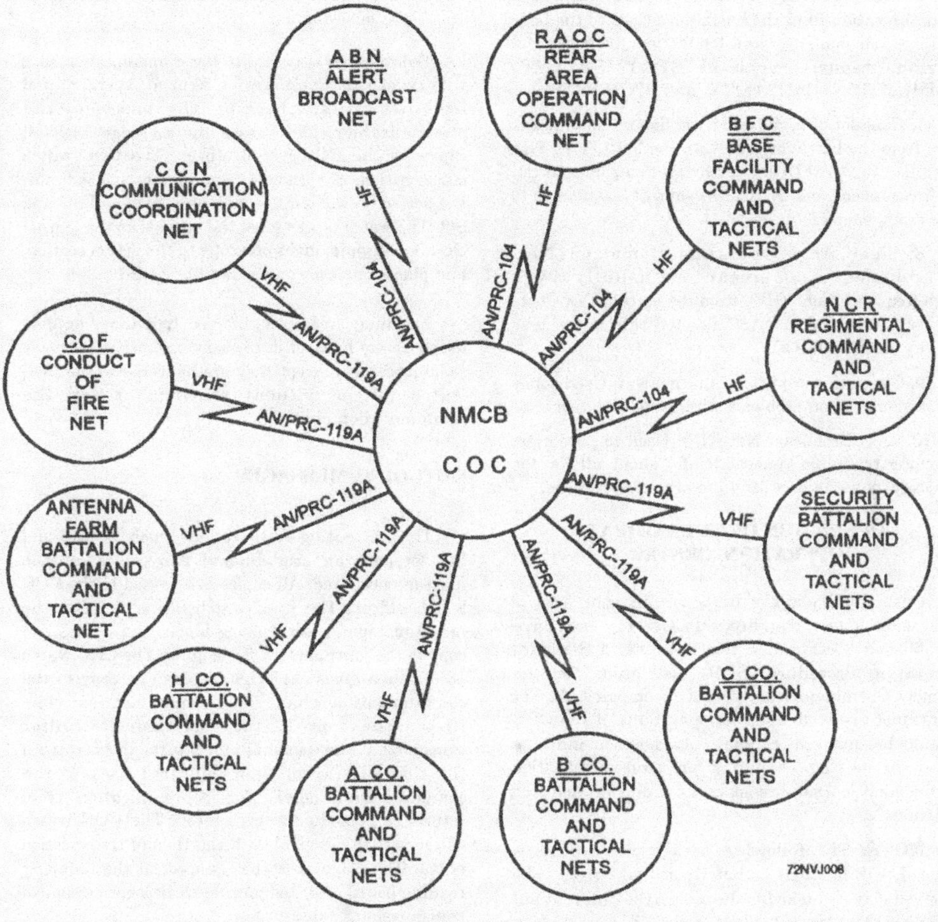

Figure 1-10.—COC radio plan.

2. Battalion Tactical Net: Establishes communication to the companies for tactical purposes and various reports, such as SPOT, SALUTE, AMMUNITION, CAUSALITY, and SITREPS.

3. Regimental Command Net: Establishes communication from the battalion's COC to the regiment when the battalion is OPCON to the regiment. Used for administrative and logistics issues.

4. Regimental Tactical Net: Establishes communication from the battalion's COC to the regiment when the battalion is OPCON to the regiment. Used for tactical purposes and various reports, such as SPOT, SALUTE, AMMUNITION, CAUSALTY, and SITREPs.

5. Base Facility Command Net: Establishes communication from the battalion's COC to the base facility commander. Used for administrative and logistics issues.

6. Base Facility Tactical Net: Establishes communication from the battalion's COC to the base facility commander. Used for tactical purposes and various reports, such as SPOT, SALUTE, AMMUMTION, CAUSALITY, and SITREPS.

7. Conduct of Fire Net: Establishes communication from the battalion's COC to the battalion's Fire Direction Center (FDC). Used by the companies to call in fire missions and by the fire support coordinator to call in fire support.

8. Rear Area Operating Command Net: Establishes communication, usually by a high-frequency radio (HF), from the battalion's COC to higher authority (i.e., RASC) located outside the base facility, such as RAOC.

9. Communication Coordination Net: Used solely for communication problems within the battalion.

10. Alert/Broadcast Net (HF): Used to pass alert warning traffic or general traffic about all (or the majority) of the units within the area of operations.

OPERATION OF THE COMBAT OPERATION CENTER

Once you know the units within your area of operations, who you are OPCON to, and have established a communication network, a Standard Operating Procedure (SOP) instruction for the Combat Operation Center should be prepared. All key personnel involved with the operations of the COC should be involved. Special consideration must be given to message handling procedures, situation boards, unit journal, fire support coordinator, and staff briefings.

NOTE: The following paragraphs contain a typical SOP for the Combat Operation Center. The one you may be tasked to design maybe different but should contain the same basic topics.

INCOMING MESSAGES

All incoming radio traffic will be written down on authorized message pads with an original plus two copies (fig. 1-11).

A guide for handling messages is as follows:

- **FLASH:** Not fixed. Handle as fast as humanly possible; objective is less than 10 minutes.
- **IMMEDIATE:** 30 minutes-1 hour.
- **PRIORITY:** 1-3 hours.
- **ROUTINE:** 3-4 hours or start of business on the following day.

The message is passed to the communication chief who checks the message for form at, spelling, and legibility. After checking the message, the communication chief passes the message with all copies to the COC watch officer. The COC watch officer assigns the action as appropriate and indicates the action section on the message, such as FSC and S-2. The watch officer gives the original to the journal clerk for logging, gives one copy to the action section, and places one copy on the reading board.

Incoming messages received by runner or field telephone are handled in the same manner as incoming radio messages, except they are delivered to the COC watch officer without routing through the communication chief.

OUTGOING MESSAGES

The originating staff sections, such as FSC and S-2, prepare an original and two copies of each message prepared. All copies are passed to the COC watch officer. The COC watch officer reviews the message, signs the releasing block, and passes one copy to the journal clerk for logging. The COC watch officer then gives the original plus one copy to the communication chief. The communication chief passes one copy to the communicator. After transmission, the communicator returns this copy with the time of transmission indicated back to the communication chief. The communication chief returns a copy to the originator. The COC watch officer gets the original with the time of transmission noted. The original is then placed on the outgoing reading board after the journal clerk logs the time of transmission.

SITUATION BOARDS

Each cognizant staff section, such as Ops and INTEL, is responsible for maintaining current situation maps and other tactical information aids. Close monitoring of the developing tactical situation is required to ensure that request for support or information are provided expeditiously. The COC watch officer should ensure that any information requiring the updates of maps or charts is forwarded to the staff sections rapidly.

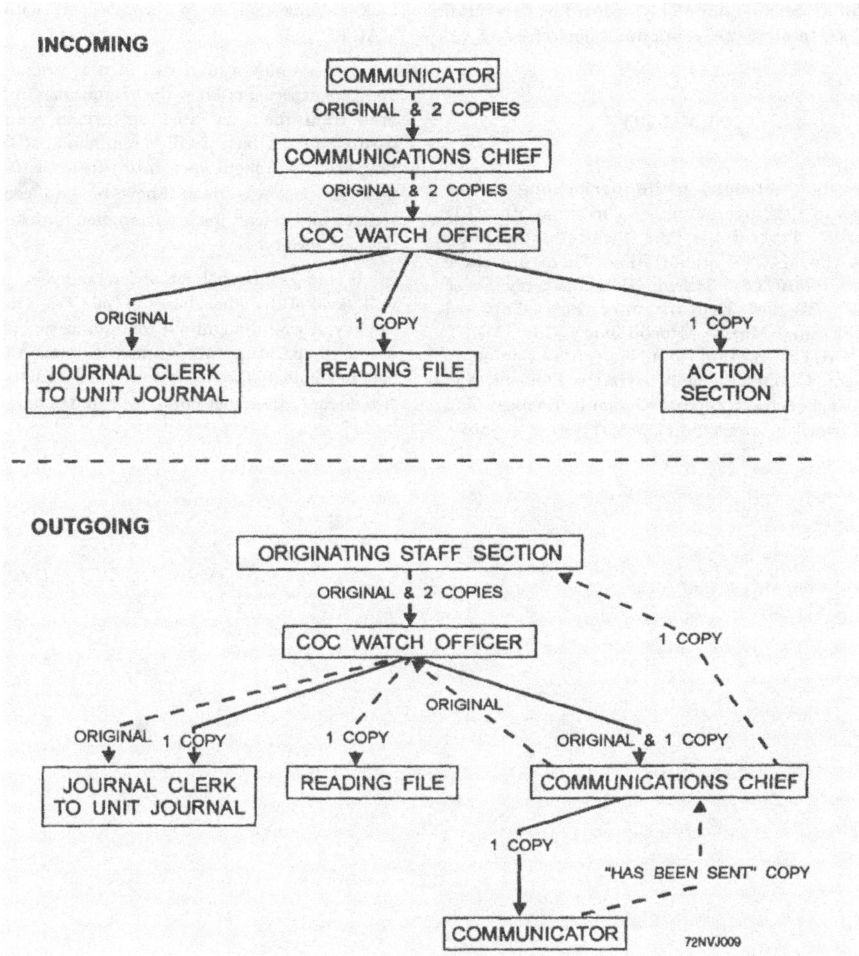

Figure 1-11.—COC message handling routing.

UNIT JOURNAL

The unit journal provides a summary of activities that is used by commanders and oncoming duty personnel to determine the current tactical situation. It covers a 24-hour period beginning at 0001 local time. The journal clerk prepares the journal under the direction of the COC watch officer. All items will be entered in a brief, accurate form. Periodic reports, such as SITREPs, will be referenced only.

FIRE SUPPORT COORDINATOR

This activity is included in the COC only when the battalions are using their mortars or providing liaison with higher commands for supporting arms. The FSC maintains maps, charts, and target lists for use during the defense.

STAFF BRIEFINGS

The operations officer (S-3) coordinates briefings that present the current tactical situation to the battalion commander. The order for this staff brief is Intel (S-2), communication chief (S-3), and FSC.

SUMMARY

To be proficient while performing duties related to COC operations, you must see the "Big Picture." Part of the "Big Picture" is to know about the MAGTF organization. There are three types of MAGTFs: Marine Expeditionary Force (MEF), Marine Expeditionary Force-Forward (MEF-F), and Marine Expeditionary Unit (MEU). All MAGTFs contain four elements: Command Element, Ground Element, Aviation Element, and Combat Service Support Element. Seabees can be attached to any type of MAGTF as a separate task organization or to any element within the MAGTF.

Seabees play a vital role in rear area security. They are expected to know the organization in the rear area from the rear area operations commander (RAOC) to the base facility commander (BF). The rear area is broken down into clusters called base facilities. Seabees must know how to command, control, and defend the area assigned to them within the base facility.

It is important that you understand your duties as well as all of the other duties in the COC. Training is the key. A well-thought-out communication plan and standard operating procedures within the COC will help to eliminate the "fog of war." As our motto states: "we do more then just build; we can defend what we build."

CHAPTER 2

ORGANIZATION AND OPERATION OF THE COMPANY COMMAND POST

In the previous chapter, we learned about the importance of a proficient Combat Operations Center (COC). For a Combat Operations Center to be effective, it must receive accurate information immediately from the companies assigned to the unit. The line companies (Alfa, Bravo, Charlie) man the defensive lines that are the battalion's portion of the defensive perimeter in the rear area or the Forward Edge of the Battle Area (FEBA); therefore, they are the **"eyes and ears"** of the COC. Company command posts are established to help the companies maintain command and control of their perspective companies. This chapter provides detailed information about the company command post and how the data it provides to the COC is vital to an effective command and control system.

COMPANY COMMAND POST

The company command post (CP) is the central point from which company operations are directed. The CP is established to provide the company commander centralized command and control facilities for the platoons assigned to the company. The platoon commanders report all activities to the company CP, regardless of their magnitudes, concerning their perspective platoons. The company CP reports to the COC all information concerning the company.

The specific composition and functions of the company command post vary with the mission assigned to the company. The normal functions of the company CP are similar to the COC and include the following:

1. Receiving and recording operational reports from the COC.

2. Maintaining current maps and overlays of friendly and enemy situations within the company's area of responsibility. This information is displayed within the company command post.

3. Maintaining current maps and overlays of the company's patrol routes and other routes that may come in contact with the company's area of responsibility.

4. Preparing and submitting operational reports to the COC concering the company.

5. Providing dedicated communication channels for tactical and nontactical operations to the COC and to the platoon commanders.

6. Transmitting orders and tactical decisions of the battalion's COC to the platoons as required.

7. Monitoring the progress of the battalion's tactical operations and reporting immediately to the COC any significant event or incident concerning the company's area of responsibility.

LOCATION

The location considerations of the company CP are similar to the location considerations of the COC. The company CP must be hard to detect from enemy forces and must be easy to defend and easily accessible to battalion personnel. Other factors to consider am centralization, traffic, communications, and defense security.

Centralization

The company command post is located rear of the forward platoons manning the defensive perimeter or the FEBA and is centered among those platoons to enhance the execution of command and control operations.

Traffic

All of the traffic entering the company CP is by foot. Personnel approaching and returning to their fighting positions from the company CP should do so by alternate routes. When this rule is not followed, a beaten path leading to the company CP for the enemy to follow will result.

Communication

Communication is vital to the company CP for the effective exercise of command and control. The company CP must have constant communication with the platoon commanders, the listening posts (LP), and the observation posts (OP). The platoon commanders must have constant communication with the troops on the front lines. A break in communication can cause massive confusion and could result in heavy casualties and the loss of lives. Communications is covered in depth farther on in this chapter.

Defense and Security

Defensive and security measures for the company CP are similar to that of the COC. The major exception is that the company CP is not as fortified as the COC. For example, barbed and tactical wire entanglements are not used around the company CP because of the constant flow of traffic entering the CP. The company commander is concerned principally with the defense of the company's area of responsibility.

ORGANIZATION OF THE COMPANY COMMAND POST

The internal organization of the company CP varies with each company or contingency operation. Written Standard Operating Procedures (SOP) are established to avoid any confusion. Personnel duties, maps, and various boards are usually standard within any company CP.

The company commander is responsible for coordinating, organizing, operating, and training watch standing personnel for the company CP. The number of personnel assigned to the watch stations, which are manned on a 24-hour basis, is normally maintained at the minimum needed to meet operational requirements. The watch personnel and their duties are as follows:

Watch Chief

1. Before assuming watch, the watch chief should read the message board, be briefed on the current situation from the current watch chief, and then make liaison with the COC.

2. During watch, keep current on the tactical situation of the battalion, make routine decisions concerning the company, and notify the company commander of incidents of an unusual nature.

3. Read all incoming and outgoing messages. Take the appropriate action on messages received from the COC. Ensure the COC immediately receives accurate information pertaining to the company's area of responsibility.

4. Ensure incoming and outgoing messages follow established routing procedures.

5. Retain releasing authority on all outgoing messages.

6. Provide complete supervision of operations in the company CP.

Watch Petty Officer

1. Handle and process information relating to intelligence.

2. Gather and distribute information gained from intelligence sources.

3. Maintain overlays of the enemy situation map pertaining to the company's area of responsibility.

4. Promptly inform the watch chief of significant or unusual incidents.

5. Maintain overlays of the company's patrols and convoys.

6. Supervise the action of the communication personnel.

7. Verify correct communication security measures are being used by all assigned communicators.

Communicator

1. Monitor radio nets as assigned.

2. Use all necessary measures to minimize interception or jamming of transmission.

Messengers/Security

1. Perform all duties as assigned.

2. Ensure only authorized personnel enter the company CP.

CHARTS AND MAPS

The types of charts and maps required within the company CP vary according to the mission and the company commander. All boards and maps must be visible to all personnel in the company CP. This arrangement enables all personnel standing watch in the company CP to visualize current situations and to help them to respond to any incoming messages quickly and accurately.

NOTE: Also posted in the company CP is a clipboard that serves as an incoming/outgoing message reading file.

AREA MAP

Unlike the COC, there is only one map to maintain in the company CP. This map displays the "big picture" of the companies with emphasis on each company's fire plan and area of responsibility. The area map includes the following:

1. Friendly troop positions in the company's area of responsibility

2. Detailed company fire plan that includes claymore mines, tactical entanglements, listening and observation posts, and early warning devices such as trip flares.

3. On call targets with the designator of the target established by the COC

4. Current location of the COC

5. Location and status of patrols or convoys

6. Landing zone (LZ) locations

7. NBC corridors/fallouts

8. Overlay of known enemy positions

CHARTS

The charts posted in the CP give personnel a quick picture of the current situation. Items that need to be displayed on charts are not limited but should at least include the following:

- Company strength
- Report status of reports sent and in process
- Equipment status
- Call signs and frequencies
- Convoy status
- Food/water/ammo status of the company
- Enemy equipment/weapons
- Weather and astronomical data
- NBC corridors/fallouts
- Challenge and passwords including the alternative for the next 24 hours
- Current MOPP level

COMMUNICATION PLAN

Like the COC, a company communication plan needs to be prepared. Communication is essential for the company commander to pass information between the troops and the COC. An uncomplicated, reliable, flexible, and responsive communication plan will enable the company commander to carry out operational plans generated from the COC effectively.

The communication officer is responsible for the units overall communication plans. The company commanders work with the communication officer for the development of a company communication plan and communication assets needed by the companies. The company commander should include in the plan communication requirements for the company and a communication network.

COMMUNICATION REQUIREMENTS

Communication requirements within the company CP vary according to the mission, but the communication equipment used to fulfill the requirements is the sane. Communication equipment includes the following:

1. AN/PRC-77 or AN/PRC-119A: Radios used to communicate with the COC, the convoys, the forward observers, the listening posts, and the patrols.

2. TA-312/PT (Field Phone): A direct phone line to the COC, to the platoon commanders, and from the platoon commanders to the troops manning the lines. Usually a TA-312/PT is run from the platoon commander's fighting position to the squad leader's fighting position or a crew-served weapon fighting position on the lines.

3. SB-22/PT: The switchboard used to connect numerous TA-312s.

COMMUNICATION NETWORK

A communication network is established similar to the one covered in chapter 1. The platoon commander should also submit a platoon wire net and radio net plan. The network should have a graphical display of all communication wire linkups and radio nets monitored by the company and platoons,

NOTE: Figures 2-1 and 2-2 are wire and radio plans pertaining to an ideal situation. The one you will be tasked to design should be tailored according to the assets available, the local terrain, and the area of responsibility assigned to the company.

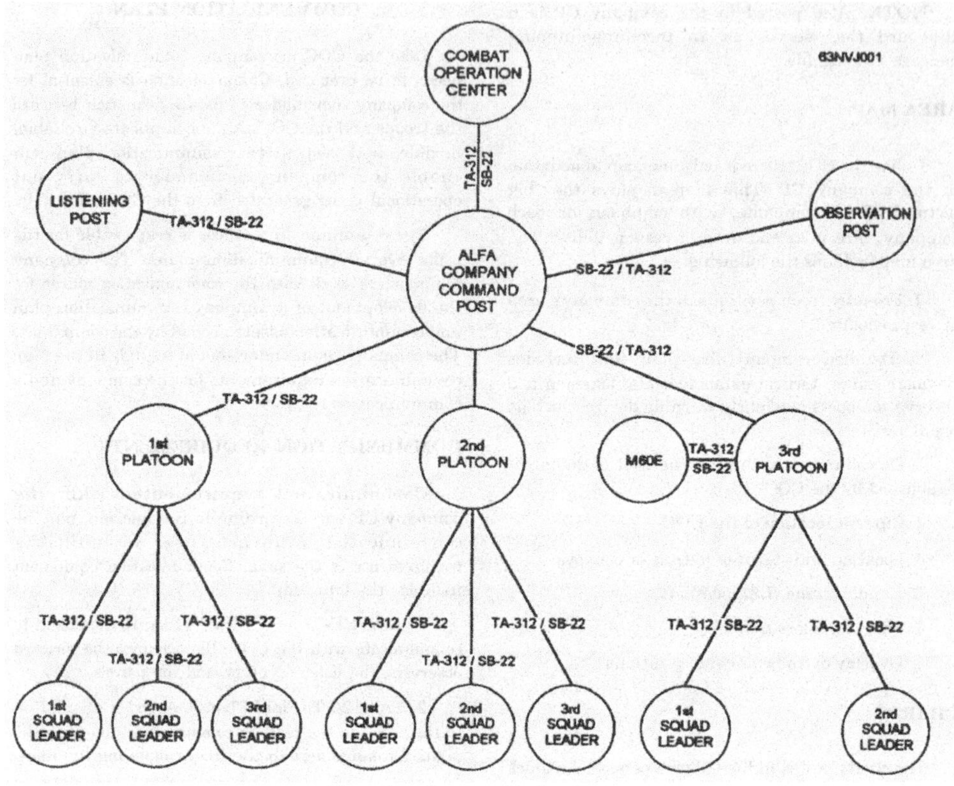

Figure 2-1.—Company CP wire plan.

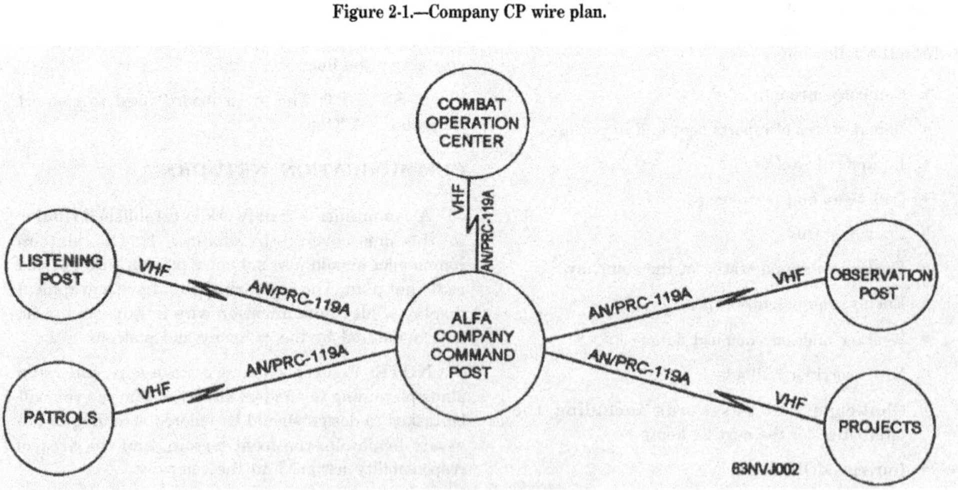

Figure 2-2.—Company CP radio plan.

The company commander should receive a copy of the Communication Electronic Operating Instruction (CEOI) to find out the radio nets the company is required to monitor. The contents of a CEOI are discussed in chapter 1. There are normally only two frequency nets that a company CP is required to monitor:

1. Battalion Command Net One: Establish communication from the companies to the COC for administrative and logistics issues.

2. Battalion Tactical Net Two: Establish communication from the companies to the COC for tactical purposes. Used for various reports, such as SPOT, SALUTE, AMMUNITION, CAUSALITY, and SITREPS.

OPERATION OF THE COMPANY COMMAND POST

A Standard Operating Procedure (SOP) instruction for the company CP should be prepared. All key personnel involved with the operations of the company CP should participate in the preparation of the SOP. Special consideration must be given to message-handling procedures and situation boards.

The following paragraphs contain a typical SOP for the company CP. The one you may be tasked to design may be different but should contain the same basic topics.

SOP FOR INCOMING MESSAGES

All incoming radio traffic should be written down on authorized message pads that allow the preparation of an original plus two copies.

NOTE: A guideline for handling messages is in chapter 1.

The communicator passes an incoming message to the watch petty officer who checks the message for format, spelling, and legibility. After checking the message, the watch petty officer should pass the message with all copies to the watch chief who will decide the appropriate action to take. The original plus one copy is then filed and one copy is placed on the reading board.

Incoming messages received by the runner or by the field telephone are handled in the same manner as incoming radio messages. When the watch petty officer is preoccupied with other events, messages should be routed directly to the watch chief.

SOP FOR OUTGOING MESSAGES

The watch petty officer prepares one original and two copies of all outgoing messages. All copies are passed to the watch chief. The watch chief reviews the message, signs the releasing block, files one copy, and forwards two copies to the communicator. The communicator logs one copy and transmits the message. After transmission, the communicator returns the original copy, with the time of transmission shown, back to the watch petty officer. The watch chief receives the original with the time of transmission noted. The original is placed on the outgoing reading board.

NOTE: When the watch petty officer is preoccupied, the watch chief prepares outgoing messages.

REPORTING PROCEDURES

Reporting procedures must be established for the company CP to be effective. The use of standardized reports speeds information flow and simplifies support requests. Proper communication procedures, including encryption of sensitive information, must be followed in submitting all reports. Every person in the company must know the proper reporting procedures. Information from the lines must get relayed back to the company CP in an accurate and timely fashion. Formats for reports may vary from unit to unit. The two most commons are the SALUTE and SPOT reports.

The platoon commander receives information from the lines by landline or radio. The primary means of communication between the platoon commander and squad leaders are wire. Messengers, visual signals, personal contact, or whistles may be used when they are more appropriate than phones and radios.

As previously mentioned, the location of field phones on the lines is dictated by the assets available, the terrain, and the size of the area of responsibility assigned to the company. The platoon commander must be able to relay information quickly and accurately. The platoon commander must also have complete command and control of the squads and crew-served weapons assigned to the platoon.

The ideal situation is for the platoon commander to have a TA-312/PT (field phone) at each squad leader's fighting position or crew-served weapon position. The field phones will tie into a switchboard

(SB-22/PT) located in the platoon commander's fighting position. The platoon commander will be tied into the company CP. Figure 2-3 shows an ideal radio network.

The platoon commander relays the information to the company CP, using the acronym SALUTE (Size, Activity, Location, Unit, Time, Equipment). The company CP relays the information to the COC with a SALUTE report.

A SPOT report is a hastily modified SALUTE report containing less detail. An example of the difference between the two reports is when the

Figure 2-3.—Various platoon and radio net illustration.

company is engaged in a fire fight on the lines. The platoon commander must concentrate on the fire fight and has little time to send a formal SALUTE report to the company CP. Several SPOT reports are used to maintain communication with the company CP. The company CP is also busy concentrating on the fire fight and will relay several SPOT reports to the COC. When the fire fight is over, the company CP will send the COC a formal SALUTE report based from the SPOT reports.

A SALUTE report is used when the observed activities of the enemy do not pose a threat to the company. This gives the platoon commander and the company CP time to evaluate the situation and to send the COC a formal SALUTE report.

The important thing is to keep the COC informed of everything and maintain the top priority of defending the company's area of responsibility. **What may not seem like vital information to you may be vital information to someone else.** When in doubt send a SPOT or SALUTE report.

AREA MAPS AND CHARTS

The watch petty officer is primarily responsible for maintaining the situation map and other tactical information aids; however, this task is usually shared with the watch chief. The watch chief ensures that any information requiring the updates of maps or charts is quickly carried out.

SUMMARY

As you can see from this chapter, the company CP plays a vital role in relation to the COC and the overall defense plans of the battalion. The organization of the company CP is similar to the COC but on a smaller scale. The primary difference is the communication from the defensive lines or the FEBA to the company CP. Emphasis is placed on defending the company's area of responsibility while keeping the COC informed. A SPOT report is a rapid way to inform the COC of vital information and should be followed up with a SALUTE report. The COC depends on the company CP for accurate and concise information. Strong coordination between the COC and the company CP can ensure a victorious outcome for any combat mission assigned to a Seabee organization.

CHAPTER 3

SETUP AND CONTROL OF MEDICAL EVACUATION (MEDEVAC)

Medical evacuation and casualty care are responsibilities shared by everyone involved with command and control activities. All personnel assigned to the unit must be aware of casualty facilities and medical evacuation procedures (MEDEVAC). A well-developed MEDEVAC plan will not only save lives but will also ensure the unit is ready for present and future missions. The loss of any Seabee because of a poor MEDEVAC plan is inexcusable and can result in a disastrous lack of firepower when it is most needed. This chapter covers in depth the setting up of a landing zone, the landing zone brief, the medical evacuation request, and the medical evacuation procedures.

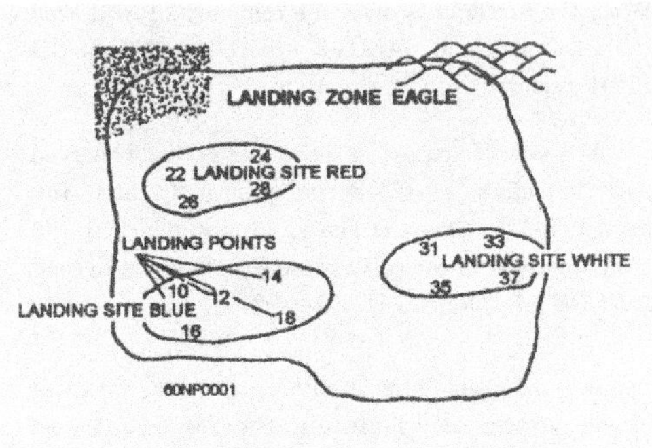

Figure 3-1.—Helicopter landing zone (LZ).

LANDING ZONE/SITE/POINTS

A helicopter landing zone (LZ) is a specified ground for landing helicopters to embark or disembark troops or cargo. A landing zone is designated by a code name. It may include one or more landing sites.

Depending upon the terrain and the size of the Seabee unit, you can divide the LZ into several **landing sites.** A landing site is a specific location within a landing zone in which a single flight of helicopters may land to embark or disembark troops or cargo. Landing sites are designated by color, such as landing site red. A landing site contains one or more landing points (fig. 3-1).

A helicopter landing point is an area within a landing site where an individual helicopter can land. Landing points are designated by two-digit numbers, such as landing point 12. For pathfinder purposes, the landing points are identified by the use of smoke or air panels. For night operations, you can mark the landing points with some type of low-intensity light. A general ***rule*** is to position landing points ten times as far from an obstacle as the obstacle is high (10:1 ratio).

NOTE: In most cases, a Seabee unit will be required to construct a landing zone with one landing site and one landing point for resupplies, troop movement, or medical evacuations.

PREPARATION OF THE LANDING ZONE

When planning the preparation of an LZ, you should take several factors into consideration. First, you should know what type of helicopters will be using the landing zone. The Combat Operation Center (COC) can provide this type of information. Second, you must consider the Seabee unit's position in relation to the enemy. Security troops must establish a 360-degree perimeter around the landing zone to defend the LZ. A third factor is the time it will take to prepare the landing zone. And a fourth factor considered is the equipment needed to prepare the LZ.

Approaches and Exits

The ground approaches to the LZ and exits from the LZ must be free of major obstacles that might obstruct landing or takeoffs, such as tall trees, telephone poles, or power lines. Approaches and exits should also be clear of obstructions that are 10 meters or higher, extending at least 50 meters in the direction of approach and exit paths. The rule of thumb for determining the distance required between the landing point and a high obstruction is a 10:1 ratio. This means that the distance a landing point is located from a tree is ten times the height of the tree.

Example: A helicopter landing or taking off near a 30-foot tree needs at least 300 feet of horizontal clearance (fig. 3-2).

Ground Obstacles

Obstacles on the ground, such as stumps or rocks, should not exceed 1 foot in height on level ground and should be less on sloping ground.

Gradient (Slopes)

Ground slope has a considerable effect on selecting a landing site or landing point within the LZ. A helicopter cannot land safely in locations where the ground slopes more than 14 degrees. When pilots land on a slope, they prefer to land uphill because of the tail down attitude of the helicopter.

Surface Conditions

Mud, excessive dust, and loose debris are considered undesirable surface conditions for helicopters. Mud causes a helicopter to become bogged down. Excessive dust reduces visibility and compromises the location of the site. Loose debris is dangerous because they are sucked up into the rotor blades or turbine intakes, causing serious damage. Shallow water, less than 18 inches deep and with a firm bottom, can be used as a landing site.

Winds

When the wind at ground level exceeds 10 knots, the helicopter must land into the wind.

LANDING SITE DIMENSIONS

Landing site dimensions vary, depending on the number of landing points required. For each landing point, a fuselage safe circle is cleared of all obstacles, such as stumps, rocks, or bushes. Clear a rotor safe circle of all obstacles that could obstruct the rotor blades (fig. 3-3).

When there is to be more than one landing point within the landing site, separate the landing points so the helicopters can simultaneously land safely in the landing site. Use figure 3-4 as a guide in selecting the appropriate size landing zone.

MARKING THE LANDING ZONE

Once you have established the LZ, the landing sites, and the landing points, you need to direct the helicopter to the location of the LZ. The proper marking of the LZ will aid the pilot in locating it.

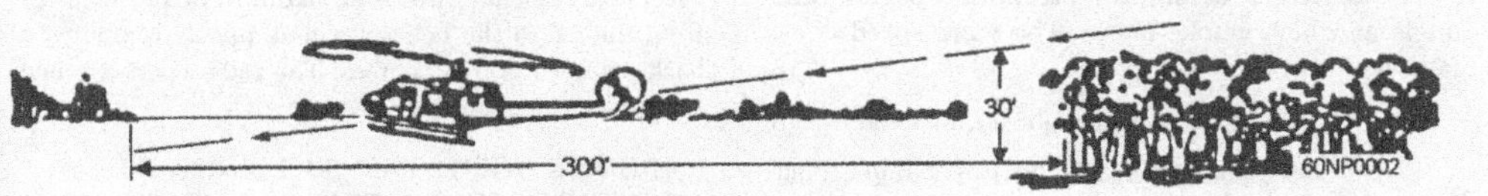

Figure 3-2.—Horizontal clearance.

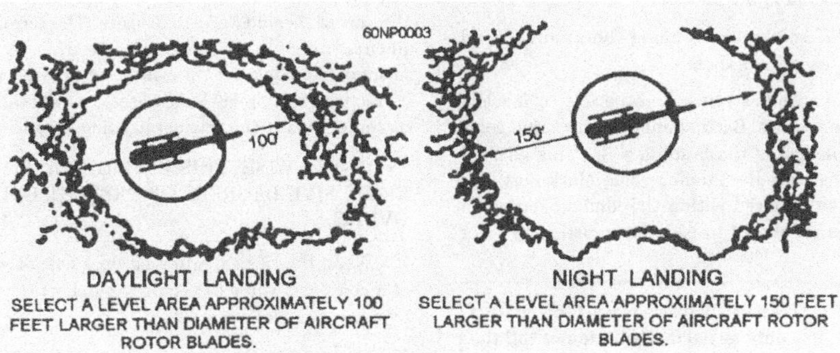

Figure 3-3.—Landing point dimensions.

TYPE	OVERALL LENGTH (FEET)	LANDING ZONE DIAMETER OBSTRUCTION HEIGHT (FLEET) 5-40	40-80	80-
UH-1E/N	57/57	100	150	200
AH-1G/J	53/53	100	150	200
AH-1T	58	100	150	200
CH-46	84	175	250	350
CH-53	89*	175	250	350

*CH-53E length is 99

Figure 3-4.—Landing zone diameter.

Recommendations for marking an LZ and for guiding a helicopter to an LZ are as follows:

1. **Daylight landing:** The landing zone is equipped with a means of showing wind direction and velocity. This is usually accomplished by the use of smoke or by verbal radio message. Expedient methods for determining wind direction and velocity are as follows:

a. **Grass drop method.** Extend your arm straight out and drop the grass from your hand. Point the extended arm at the dropped grass on the ground. The angle between the arm and the body divided by four is the wind velocity in knots.

b. **Angle of smoke method.** Observe the angle at which smoke blows. The wind speed is as follows:

—If smoke goes straight up, no wind.

—If smoke blows at a 30-degree angle, wind is 3-5 knots.

—If smoke blows at a 60-degree angle, wind is 5-7 knots.

—If smoke blows along the ground, wind exceeds 8 knots.

Use smoke and landing zone panels to mark a landing zone by day. Both should be the same color as the designation of the landing zone. This will aid the pilot in locating the landing zone, Mark obstacles that cannot be removed within the landing zone with single red panels staked to prevent uprooting by rotor wash.

SECURITY CAUTION: If smoke is used to mark the landing zone, **use only as needed and do not tell the pilot the color of smoke;** ask the pilot to acknowledge the color after the smoke grenade is set off.

2. **Night landing:** The organization and use of an LZ at night or during periods of low visibility is more complex compared to daytime operations. Special lighting equipment or field expedients as required.

a. You must indicate outlines of landing zones by low-intensity markers.

b. You must show obstacles near the landing zone by low-intensity markers or voice radio instructions.

Another method of guiding the aircraft to the zone is vector instructions. This is simply relaying instructions to the pilot by radio. For example, the radio operator spots the helicopter. Using a compass, the radio operator shoots an azimuth of 135 degrees from the LZ to the helicopter and quickly computes a back azimuth of 315 degrees. The radio operator then transmits the following message:

"HOME WISH, THIS IS FLIGHT BEE . . . VECTOR THREE ONE FIVE TO LANDING ZONE HAWK . . . OVER."

The pilot then acknowledges the message and takes up the correct heading of 315 degrees. The term *vector* is always used in a situation like this to prevent misunderstanding. As the helicopter approaches, minor corrections will probably be necessary. These are given as corrections to the original heading b y the following:

"HOME WISH, THIS IS FLIGHT BEE. . . COME RIGHT FIVE DEGREES OF PRESENT COURSE. . . OVER."

When the tactical situation does not allow the use of a compass, you can vector the helicopter to an LZ by using the clock system:

"HOME WISH, THIS IS FLIGHT BEE . . . MY POSITION IS AT YOUR NINE O'CLOCK . . . OVER."

The aircraft would then execute a 90-degree turn to the left.

LANDING SIGNALMAN ENLISTED (LSE) SIGNALS

Once the pilot has located the LZ, you can now help the pilot land the helicopter. These signals are visual arm-and-hand signals used by personnel to "talk" to the pilot of the landing helicopter. The confidence of the pilot in the LSE's signals depends on the precise manner in which the LSE gives the signal. Movements are sharp and precise. LSE's signals are executed exactly as prescribed. The LSE must always remain alert for signals from the pilot. During night operations, LSEs must use illuminated wands. During a landing approach, the LSE's functions areas follows:

1. Inform the pilot of the approaching helicopter that you are the LSE. You do this by means of the Prepare for Guidance signal.

2. Indicate the landing point to the pilot by positioning yourself 25 meters in front of and 10 meters to the right of the landing point as the pilot looks at it.

3. Aid the pilot in landing safely on the landing point. The pilot is responsible for the approach and landing of the helicopter. However, the pilot relies heavily on the LSE to provide warning of conditions of which he or she is not aware of and to direct the helicopter to a safe landing point.

4. The only signal that the LSE must give to the pilot on the deck is the Wave-Off signal. All other signals are advisory signals to the pilot. The decision to acceptor reject the signal is solely the responsibility of the pilot. Responsibility for the safety of the aircraft can never be relinquished to the LSE. The Wave-Off signal is given when it is not safe for the helicopter to land.

5. Indicate to the pilot when it is safe to take off.

NOTE: Figures 3-5 through 3-14 are Initial Terminal Guidance LSE signals that are standard throughout the Marine Corps. Any misunderstandings that may arise about signals are clarified by close coordination with the supporting helicopter unit. The signals are used for daytime operations. Signals given at night are executed in the same manner except that an illuminated amber director's wand (Chem Lite) is held in each hand.

Figure 3-5.—Prepare for guidance.

Figure 3-6.—Forward.

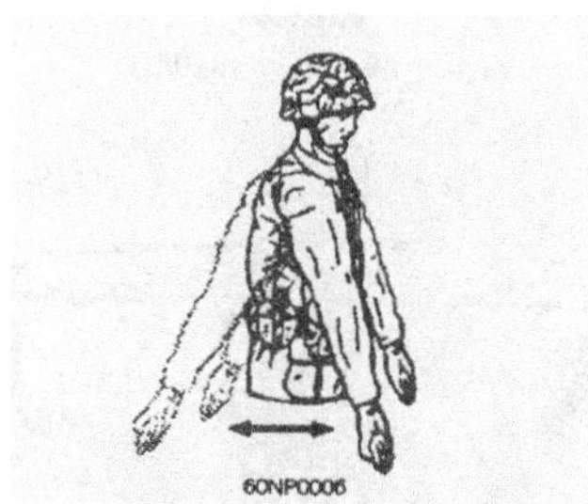

Figure 3-7.—Backward.

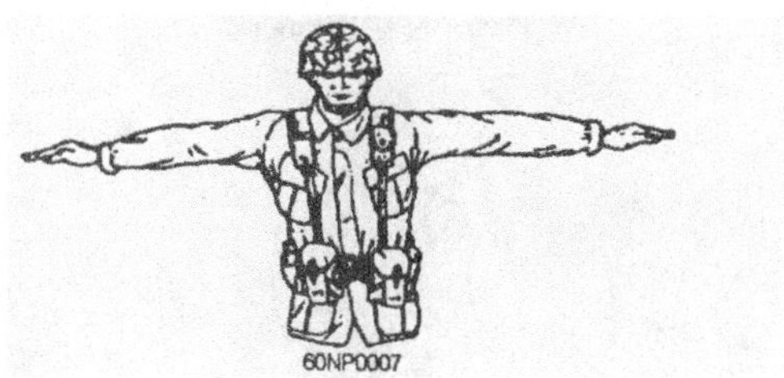

Figure 3-8.—Hover.

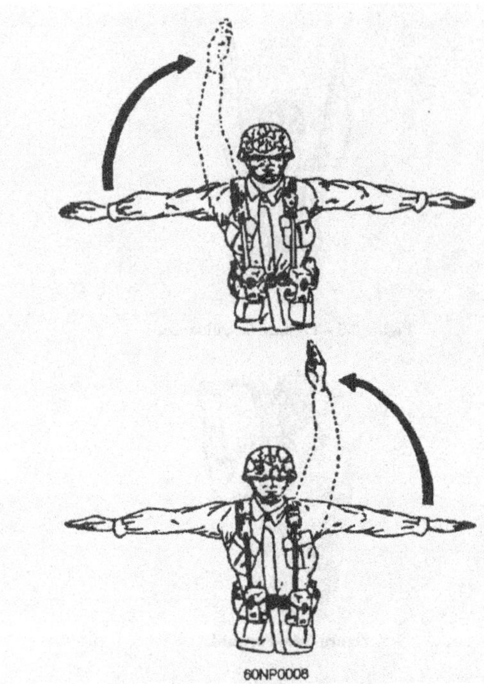

Figure 3-9.—Move right/left.

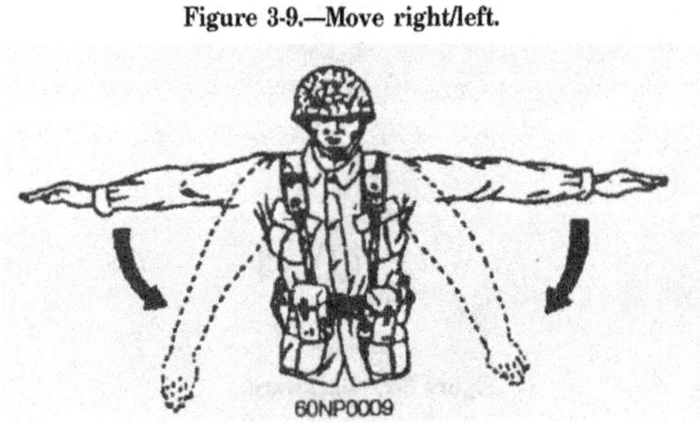

Figure 3-10.—Downward.

Figure 3-11.—Land.

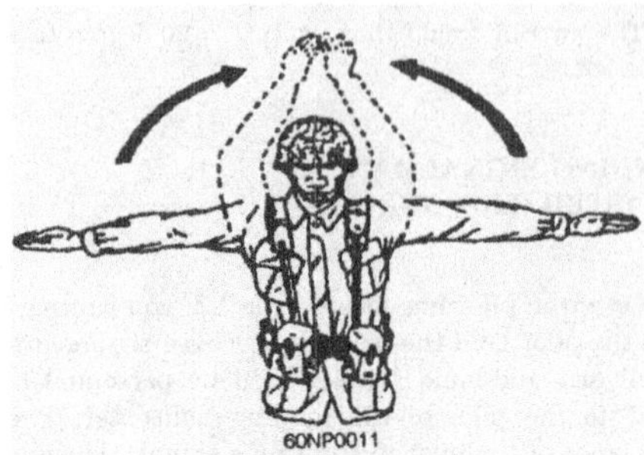

Figure 3-12.—Upward.

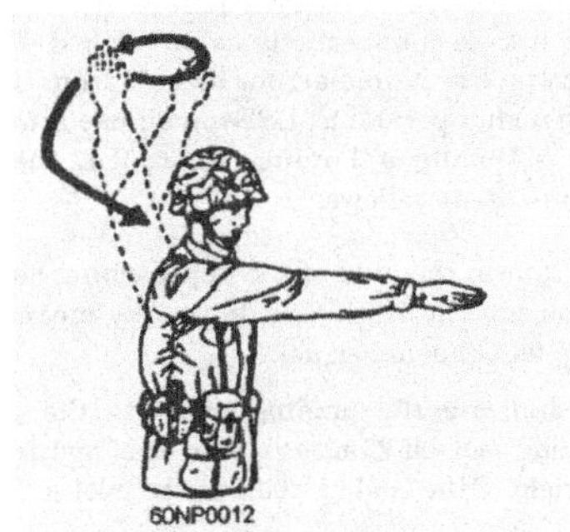

Figure 3-13.—Takeoff.

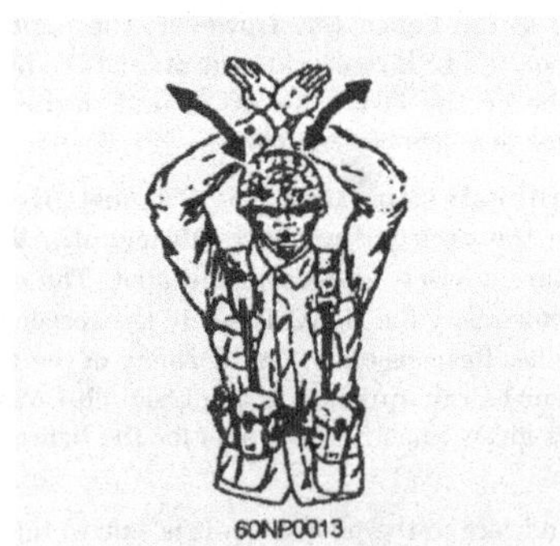

Figure 3-14.—Wave-off.

LANDING ZONE BRIEF

When a helicopter has to land for troop pickup/drop, resupply, MEDEVAC, or for any other reason, the pilot must have certain information to ensure a safe landing. This information is provided in a landing zone brief (fig. 3-15).

	LANDING ZONE BRIEF	
1.	MISSION NO.	______________
2.	LOCATION COOR/RAD/DME	______________
3.	UNIT CALL SIGN	______________
4.	FREQUENCY	PRI UHF ___ FM ______
		SEC UHF ___ FM ______
5.	LZ MARKING	______________
6.	WIND DIRECTION/VELOCITY	______ / ______
7.	ELEVATION/SIZE	______ / ______
8.	OBSTACLES	
9.	FRIENDLY POSITIONS: DIRECTION/DISTANCE	______ / ______
10.	ENEMY POSITIONS: DIRECTION/DISTANCE	______ / ______
11.	LAST FIRE RECIEVED: TIME/TYPE	______ / ______
12.	DIRECTION OF FIRE/DISTANCE	______ / ______
13.	CLEARANCE TO FIRE: DIRECTION/DISTANCE	______ / ______
14.	APPROACH/RETIREMENT (RECOMMENDED)	______ / ______
15.	PERSONNEL/EQUIPMENT	______ / ______
16.	OTHER	______________

60NP0014

Figure 3-15.—Landing zone brief.

Each line of the landing zone brief is explained below.

Line 1: The mission number will be assigned to the pilot by Direct Air Support Center (DASC) personnel. Under normal circumstance this line is not used by Seabee units.

Line 2: Give at least a six-digit grid to identify the location of the LZ.

Line 3: Your call sign.

Line 4: Self-explanatory.

Line 5: List the methods of marking the LZ. For example, smoke, air panels, signal mirrors, lights, and so forth.

Line 6: State the direction from which the wind is coming and the wind speed if known.

Line 7: State the elevation of the LZ (air is thinner

Line 8: An obstacle is anything higher or deeper than 1 foot on the LZ or anything near the LZ that may create a hazard to the aircraft. Explain where the obstacles are in relation to the LZ.

Line 9: State where friendly troops are in relation to the LZ.

Line 10: State the location of the enemy in relation to the LZ.

Line 11: Self-explanatory.

Line 12: Self-explanatory.

Line 13: State the direction and distance the helicopter can fire without endangering friendly troops.

Line 14: If possible, the approach heading should be into the wind. The retirement or departure should be clear of enemy positions.

Line 15: The different helicopters used for MEDEVAC operations vary in the number of personnel and equipment they can transport. Increased elevation and temperature decrease the weight capacity of a helicopter. A helicopter pilot must know exactly the number of personnel and what type of equipment is to be transported.

Line 16: Anything not previously mentioned that could help the pilot in anyway.

MEDICAL EVACUATION REQUEST (MEDEVAC)

A MEDEVAC request contains all the information that Direct Air Support Center (DASC) personnel need to dispatch a helicopter to your position. In a combat situation, the Combat Operation Center (COC) routes the request from the battalion aid station to higher headquarters. Higher headquarters will route the request to DASC personnel. In extreme situations where communication to the COC or the battalion aid station is interrupted, platoon commanders or company commandes should have the knowledge and capabilities to request a MEDEVAC. The operation plan (OPLAN) will explain the routing of a MEDEVAC request. A MEDEVAC request does not actually bring the helicopter to the LZ. The request simply identifies the requesting unit, location of the LZ, and describes the casualty. The DASC will make the necessary arrangements/decisions concerning the MEDEVAC request. The format in figure 3-17 contains all the information required in a MEDEVAC request. The format may change slightly at diffenmt organizations.

MEDEVAC PROCEDURES

The importance of a well-thought-out MEDEVAC plan cannot be emphasized enough. The lives of everyone in the unit depend on it. All personnel assigned to the unit must know the MEDEVAC procedures. The elements requiring special consideration are litter bearers, ambulance, battalion aid station, and LZ security.

LITTER BEARERS

Litter bearers should be personnel, such as messcooks, personnelmen, and yeomen, from Headquarters Company. The litter bearers are assigned as stretcher teams with the COC being the prime coordinator for their utilization. The COC will inform the battalion aid station for their assistance. Many variations occur however, and it must be emphasized that this source of litter bearers is not taken for granted. Platoon commanders and the company chief must be assured of an adequate number of litter teams. Eight men per company are considered adequate unless mass casualties are anticipated. Other sources of litter bearers are as follows:

1. Members of the platoon. This source is mentioned only to be discouraged, although it will be well recognized that in an emergency this may be the only source available. Utilization of platoon personnel manning the defensive lines or FEBA may result in a disastrous lack of fire power when it is most needed for defense.

2. Members of the reserve platoon. Here again it is recognized that in an emergency, this source must be used. However, to rely on this source is not a good practice because the reserve platoon is critically needed during a heavy attack; therefore, if the platoon is used for casualty evacuation, the company commander may be caught shorthanded.

When a casualty occurs, the wounded person, if possible, should go to a relatively protected location away from the defensive lines or FEBA. The word is then quietly passed to the CP for a corpsman or litter bearer. The CP will then inform the COC for action. Passing the word back quietly and expeditiously is important. Loud shouting for a corpsman by all the troops in the vicinity can have a demoralizing effect on other troops. Incidents have occurred when inexperienced troops have panicked because of frenzied shouting of the whole squad for a corpsman to take care of a single casualty.

Litter bearers are vital for the survival of a casualty, in the maintenance of good morale, and most of all to prevent loss of firepower for the defending platoons. For these reasons, it is vital that provisions for litter bearers be included in the MEDEVAC plan.

SAMPLE LANDING ZONE BRIEF

TRANSMISSION ITEM	YOU SAY
1. MISSION NUMBER	*(Provided by higher authority)*
2. LOCATION: COOR/RAD/DME	*Grid 456129*
3. UNIT CALL SIGN	*A4F*
4. FREQUENCY	*Primary FM 30.50, Secondary FM 45.10*
*5. LANDING ZONE MARKING	*Signal Mirror, Air Panels*
**6. WIND DIRECTION/VELOCITY	*Wind From East at 15 Knots*
7. ELEVATION/SIZE	*Elevation 2,500 Feet Size 180 Feet in Diameter*
8. OBSTACLES	*40 Foot Tree 90 Meters South of Landing Zone*
9. FRIENDLY POSITIONS: DIRECTION/DISTANCE	*Friendlies Southeast 100 Meters*
10. ENEMY POSITIONS: DIRECTION/DISTANCE	*Enemy Southeast 500 Meters*
11. LAST FIRE RECEIVED	*1800, Small Arms*
12. DIRECTION OF FIRE/ DISTANCE	*Enemy Fire From Southeast 500 Meters*
13. CLEARANCE TO FIRE; DIRECTION/DISTANCE	*South and Southeast 150 Meters From Landing Zone*
14. APPROACH/RETIREMENT (RECOMMENDED)	*Approach Heading 90° Return Heading 350°*
15. PERSONNEL/EQUIPMENT	*Six Personnel with Alice Packs & Rifles*
16. OTHER	*5° Slope*

* **CAUTION: As a security caution, if you use smoke to mark landing zone, DO NOT tell pilot what color smoke will be. Ask pilot to acknowledge color after grenade is set off.**

** **Determine and report wind condition and direction:**

For angle of smoke method, observe smoke blowing if the wind is blowing.

SMOKE STRAIGHT UP	NO WIND
SMOKE 30°	WIND AT 3-5 KNOTS
SMOKE 60°	WIND AT 5-7 KNOTS
SMOKE ALONG GROUND	WIND IN EXCESS OF 8 KNOTS

60NP0015

Figure 3-16.—Sample of landing zone brief.

MEDEVAC REQUEST

Mission # ________

1. Emergency ______ Priority ______ Routine ______

2. Requesting until call sign

3. Date-time group

4. LZ or pickup coordinates (clear)

5. Number of WIA ____ KIA ____ SICK ____ OTHER ____

6. Pickup of doctor or corpsman at LZ or Coord

7. Airborne medical assistance required yes ____ no ____

8. LZ is marked with Smoke ____ Panels ____ Light ____ Strobe ____

9. LZ Freq Designator ______ Call Sign ______

10. Remarks: Need jungle penetrator?

* Casualty report must follow this request.
DASC is informed when MEDEVAC is completed.

Figure 3-17.—MEDEVAC request.

AMBULANCE

An ambulance is normally stationed at the battalion aid station (BAS) and functions as far forward as the terrain and enemy activity permits. The ambulance is primarily used to prevent the prolonged carrying of litters. When the ambulance is called forward, explicit instructions are given as to routes taken and the exact location to which it is to go. Ambulances frequently draw enemy mortar and artillery fire; therefore, they should not be brought up to an area under direct observation of the enemy.

BATTALION AID STATION

The battalion aid station is the central location that all casualties are taken to before any type of evacuation. Medical staff personnel are the only personnel qualified in determining whether a casualty *MEDEVACed).* Other functions of the battalion aid station are to give further first aid, check for continued hemorrhage, rebandage where needed, and apply splints if needed. Seriously wounded casualties are given supportive therapy for shock in the form of plasma and serum albumen in preparation of a MEDEVAC. Once the medical staff has determined the priority of the casualties requiring a MEDEVAC and those not expected to survive the flight, the COC is contacted and sends a MEDEVAC request to higher authority. It is recommended that the chaplin is available for those not expected to survive.

SECURITY

Once the MEDEVAC request has been sent to higher authority, the COC will quickly activate the security team. The security team (React Force) is responsible for securing the LZ, establishing the helicopter. Also controlled by the security team is the staging area for those requiring MEDVAC. A member of the security team is established as a "Pit Boss" that will control the litter bearers and all movement of the wounded to the LZ.

Securing the LZ

Weapons and personnel required to secure an LZ are dictated by the size and terrain of the LZ. A 360-degree perimeter must be established around the LZ. All likely avenues of approach and sectors of fire must be covered. Helicopter pilots will not land if they feel the LZ is not properly secured.

Establishing Communication

Simultaneously, communication personnel assigned to the security team will establish communication at the landing site. It is important that communication personnel know the proper mission frequency for contacting the pilot and keeping in contact with the COC. The type of communication equipment needed depends on the terrain of the LZ. Usually, two PRC 119As that are equipped with a AS-3683 10-foot whip antenna are used. Two radios are required because communication must be maintained with both the pilot and the COC.

Casualties

Once the LZ has been secured, casualties are transported from the battalion aid station to the landing point by any means necessary. The casualties are arranged in priorities and according to the capacity of the helicopter.

When en route to the LZ, the helicopter pilot should inform the communicator at the LZ of the capacity of the helicopter and if a corpsman is onboard the helicopter. For example, the pilot will inform the communicator that the helicopter has room for three litters and one walk-on. The medical staff will then arrange the casualties in that order. Casualties must be tagged with their name, rank, SSN, unit, and type of injuries. All loose articles of clothing, such as 782 gear, are removed from the casualties.

Landing and Loading the Helicopter

Once the pilot has acknowledged the location of the LZ by means discussed previously in this chapter, landing signalman enlisted signals are used to land the helicopter. Only one person is designated as the landing signalman. Once the helicopter has landed and a representative disembarks the aircraft, the landing signalman directs the representative to the senior medical staff personnel in charge. The helicopter representative will coordinate with the medical staff representative on how the casualties are to be loaded on the aircraft. Once all the casualties have been MEDEVACed, the COC will send a SALUTE and WIA (or Casualty CASREP) report to higher headquarters.

SUMMARY

Evacuation by helicopter is the ideal evacuation method. This type of evacuation is usually available day or night. However, helicopters should be used with discretion as their number is usually limited and they draw enemy fire when observed. The individual Seabee should be taught basic first aid and, if wounded, to remain calm. If the wound is minor, a buddy should apply a battle dressing and continue to deliver fire until the action lessens. The ability of a buddy to give first aid depends on the tactical situation. Properly establishing an LZ and mutes to the LZ from the BAS is important to any MEDEVAC plan. Pilots are just as concerned for their safety as you are about your troops safety. Remember, proper setup and control of MEDEVAC procedures are critical and will ensure that the individual Seabee will be back to build and fight another day.

CHAPTER 4

PLANNING AND DEVELOPMENT OF DEFENSIVE TACTICS

Seabee battalions are primarily a defensive unit during combat and must consider the planning of defensive tactics as top priority. Poor planning of defensive tactics will not only endanger the battalion but will also endanger all other units involved with the operation. Other units assigned to the theater of operation depend on Seabee units to defend their area of responsibility. Defensive operations for Seabee battalions include those actions taken for destroying or trapping a hostile force, denying an enemy access to an area and reducing the capabilities of the enemy. The goal is to accomplish these actions with minimum or no losses to the battalion. This chapter covers in depth the concept of defense, the fundamentals of defense, the types of defense, and the platoon, squad, and fire team defensive positions. Also covered are the crew-served weapons employment, the platoon fire plan, and the company fire plan. Although this chapter speaks mainly of defensive tactics for a Seabee battalion, Seabee units can also implement the context of this chapter.

CONCEPT OF DEFENSE

In the defense, the defender takes every opportunity to seize the initiative and to destroy the enemy. The defender seizes the initiative by forcing the enemy to react in conformity with the defensive plan of the battalion and exploiting the enemy weaknesses. As a platoon commander or assistant company commander, it is important for you to know how the defensive plan relates to the overall defensive operation. Seabees are required to set up a defensive perimeter or, in extreme cases, defend the Forward Edge of the Battle Area (FEBA). The defensive tactics used in each case are similar.

NOTE: For simplicity purposes the defensive perimeter or the defensive lines will be referred to as the FEBA.

DEFENSIVE AREAS

defensive areas, established at the defensive perimeter or FEBA, include the security area, the forward defense area, and the reserve area. Each of these areas is allocated forces and fires as a part of the complete defense plan (fig. 4-1).

Security Area

The battalion security area begins at the FEBA or the defensive perimeter and normally extends 500 meters to the front and to the flanks of the battalion. Depending on where battalion security elements are used, this area can be increased. Seabee forces in the security area include the listening post, the observation post, and the patrols that furnish information about the enemy; delay, deceive, and disrupt the enemy as much as possible. Division forces, operating deep in the security area (beyond 500 meters), will consist of general outposts (GOPs), combat outposts, flank security forces, division aerial surveillance elements, and patrols. As a platoon commander or patrol leader, you must be aware of all forces operating in the security area.

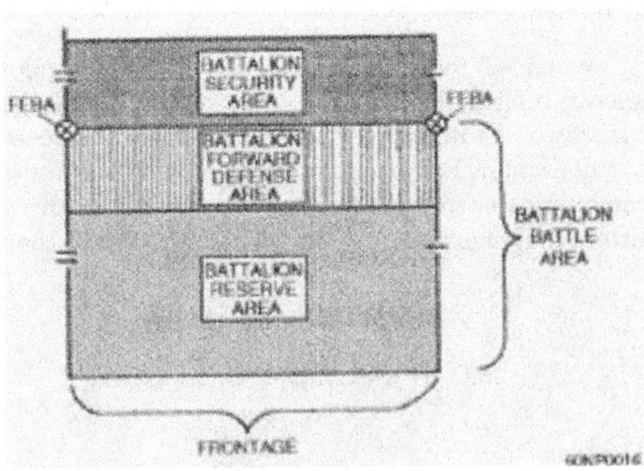

Figure 4-1.—Battalion defensive areas.

Forward Defense Area (FDA)

The forward defense area extends rearward from the defensive line or FEBA to the rear boundaries of the frontline companies that are used as the forward defense echelon of the battalion. The composition of the forward defense echelon depends upon the form of defense used.

Reserve Area

Extending from the rear boundary of the forward defense echelon (frontline companies) to the rear boundary of the defense of the battalion is the reserve area. The reserve forces and those uncommitted forces under battalion control occupy positions in the reserve area and add depth to the defensive position. The reserve is the principal means by which the battalion commander influences the defensive battle and regains the initiative.

FUNDAMENTALS OF DEFENSE

Effective defensive actions are achieved by a blend of the principles discussed below. The degree to which each principle applies will vary with the mission and the situation. This is decided only after consideration of all the principles (fig. 4-2).

PROPER USE OF TERRAIN
SECURITY
MUTUAL SUPPORT
DEFENSE IN DEPTH
ALL-AROUND DEFENSE
COORDINATED FIRE PLAN
USE OF BARRIERS
FLEXIBILITY
MAXIMUM USE OF OFFENSIVE ACTION
DISPERSION

Figure 4-2.—Fundamentals of defense.

PROPER USE OF TERRAIN

Maneuvers that are possible and dispositions that are essential are indelibly "written on the ground." Unfortunate indeed is the platoon commander or assistant company commander who is unable to read this "writing." The intelligent leader knows that terrain is a major ally and that it virtually decides the positioning of platoons and squads in the defense. The defender retains control of terrain features essential to observations, communications, and maneuver of reserves. Denying the enemy the use of terrain that might jeopardize the mission of the battalion is the major role of a defender.

To read the "writing on the ground," you can organize the analysis of weather and terrain primarily around the following set of military considerations (KOCOA):

- Key terrain
- Observation and fields of fire
- Concealment and cover
- Obstacles to movement
- Avenues of approach

These considerations are discussed in the following paragraphs. While the discussions of these considerations focus on terrain, weather and terrain are inseparable. Terrain that offers good traffic conditions when dry may be impassable when wet. A hill that provides good observation on a clear day may not provide any visibility on a rainy day or at night.

Key Terrain

Key terrain features must be considered in formulating defensive tactics. Their selection is based on the mission of the command. Tactical use of terrain often is directed at increasing the ability to apply combat power. Also considered is channeling the enemy into the beaten zone of organic weapons maintained by the battalion. The selection of key terrain varies with the following:

- Level of command
- Type of unit
- Mission of the unit

Seabees are normally a service support unit for the Marine Air-Ground Task Force (MAGTF). Service support units need roads over which to move supplies and secure areas in which to construct facilities. Aviation units need high terrain on which to set up radars and communication facilities and large, flat areas for airfields.

Observation and Fields of Fire

Observation and fields of fire are so closely related that they are considered together. Fields of fire are based on observation because a target must be seen to bring effective fire upon it.

The observation considerations are as follows:

- Weather conditions
- Time of day
- Vegetation
- Surrounding terrain

The highest terrain features are generally ideal for observation. However, during times of poor visibility, positions in low areas that the enemy must pass through may provide better observation.

The field of fire is the area a weapon or group of weapons may cover effectively with fire from a given position. When you are selecting a field of fire for a weapon, it is critical to know the capabilities of the weapon. Positioning an M60E3 machine gun with a 400-meter field of fire is poorly using the weapons capability considering the weapon has a maximum effective range 1,100 meters. **The object is to engage the enemy as far out as possible.**

Cover and Concealment

Cover and concealment is used together to provide protection from the effects of fires and observation.

1. Cover is protection from the effects of fire. Examples of ideal cover are as follows:

- Rocks
- Shell craters
- Ditches or quarries
- Buildings
- Caves
- Sunken roads
- River banks
- Walls
- Folds in the ground
- Highway fills

Areas that provide cover from direct fire may or may not protect against the effects of indirect fire.

2. Concealment is the protection from observation or surveillance both air and ground. Some examples that provide ideal concealment are as follows

- Woods
- Underbrush
- Snowdrifts
- Tall grass
- Cultivated vegetation
- Other features that deny observation

Terrain that provides concealment may or may not provide cover.

Obstacles to Movement

Obstacles are anything, including a natural or artificial terrain feature, that stops, impedes, or diverts military movement. Entanglements are set up as obstacles along the FEBA or the defensive perimeter to channel the enemy into the beaten zone of various weapons held by the defending unit.

Avenues of Approach

Avenues of approach are routes the enemy is likely to travel to reach its objective. When setting up the defense positions, the platoon commander visualizes all possible enemy avenues of approach into the area.

An evaluation of the avenues of approach, together with key terrain features, serves as the basis for positioning platoons or squads for planning fire support. The area selected for defense should afford good observation, fields of fire, and adequate cover and concealment to the defending forces. The defender applies the principles of camouflage and continuously improves camouflage throughout the defensive operation.

SECURITY

Security consists of those measures that prevent surprise, avoid annoyance, preserve freedom of action, and deny to the enemy information about our forces. Security is not only keeping watch to detect the enemy but also deceiving the enemy as to the strength of the battalion, location of the weapons, and other valuable information. Using camouflage effectively is very important to security.

MUTUAL SUPPORT

A well-developed defense plan includes mutual support from adjacent units. Seabee units are normally attached to an MAGTF and have a variety of mutual support. However, conditions, created by wide unit separation, increase the problem of achieving mutual support. Support of adjacent units by fire can be hampered by distances that exceed the ranges of the supporting units organic weapons and must be considered when seeking mutual support. Mutual support becomes critical when conducting a convoy. Normally the convoy will travel out of range from a Seabee unit's organic weapons, so mutual support from adjacent units becomes critical.

ALL-AROUND DEFENSE

Although the platoon commander positions the squads toward the expected direction of an enemy attack the platoon must be prepared to defend against an attack from any direction. The possibility of multi-direction enemy attacks, including helicopterborne or airborne attack, increases the importance of this principle. All-around defense is best achieved by early warning and the rapid shifting of platoons into supplementary fighting positions to counter a developing attack. The topic of fighting positions is discussed further in this chapter. Early warning capability provided by air defense elements of the Marine Aircraft Wing (MAW) can provide defense against helicopterborne or airborne attacks.

IN-DEPTH DEFENSE

Maximum defense in depth is required to absorb an enemy attack within the battle area. Defense in depth applies to the squad level by engaging the enemy at maximum small-arms range as it advances and continuing this fire until the enemy has been stopped. The requirement for depth is increased when the enemy is superior in mobility or has the capability of destroying or rupturing defensive positions by using nuclear weapons. Organized in-depth positions must control key terrain, block avenues of approach, reduce the defender's vulnerability to nuclear weapons, and assist the maneuver of the reserve troops.

COORDINATED FIRE PLAN

Coordination of all fire from weapons organic and supporting the battalion is considered. Included in a coordinated fire plan is naval gunfire and tactical aviation support usually provided by the Marine Air Group (MAG). Coordination between the companies is also critical. The fire of the squad forces the enemy to slowdown and deploy, thus creating a target for the mortar crew of the battalion, adjacent companies, or supporting weapons. This makes the squads a key element in a coordinated fire plan.

COORDINATED BARRIER PLAN

Using barriers, either natural or man-made, can channel, direct, restrict, or stop enemy movement. The effective use of barbed wire and concertina, claymore mines, antitank mines, and antipersonnel mines, and the effective use of the terrain add to the defense of the battalion. Natural obstacles must be exploited. The creation or exploitation of barriers must be coordinated at all echelons of the command.

FLEXIBILITY

A platoon commander must continually develop various courses of action to meet the enemy threat. Being flexible is to ask yourself, What should I do if the enemy does this?

DISPERSION

Normally in a defensive situation, squads are limited to how spread out they are over a wide area. Firing positions are assigned and should be close enough to provide interlocking fire. They should not be close enough so an enemy machine gun or mortar can wipe out an entire fire team or an entire squad. The degree of dispersion will be influenced primarily by the mission and by the following:

- Terrain
- Friendly and enemy air situation
- Mobility of opposing forces
- Enemy nuclear capability
- Responsiveness of friendly nuclear and conventional fire support

USE OF TIME AVAILABLE

Since the defender can examine the terrain in detail and plan its best use, every effort is made to prepare the defense in advance. The time available for planning and preparing for the defense will influence the following:

- Tactical employment of the companies
- Preparation of obstacles
- Coordination of fire from each company
- Coordination of supporting fire
- Priority for performance of tasks

The effectiveness of the defense depends not only on the time available for its planning and preparation but also on its advantageous use during the preparation phase. This fundamental also applies after the preparation phase because improvement continues during the defense.

POSITIONING THE SQUAD IN THE DEFENSE

The effectiveness of a unit in combat is largely dependent on selecting firing positions that allow the weapons of the unit to be used effectively and to keep the enemy from using their weapons. A primary duty as a platoon commander or squad leader for a rifle platoon is positioning the squads in the defense. When you are employing a squad in the defense, it is good to review your previous experiences as a squad member. Reflect upon the reasons you were positioned as a rifleman, grenadier, or other member of a squad to better determine what type of defensive firing position is needed. There are many points to consider in selecting a firing position. Some of these are as follows:

1. **Mission:** If the mission of the unit is to defend an airfield, the firing positions of the unit must facilitate the accomplishment of the mission, despite how good other firing positions may be.

2. **Enemy:** The capabilities of an enemy usually determine how he will attack. Mechanized units usually attack through fields and other open terrain where they can use the speed of their vehicles to an advantage. Infantry units, however, seek to use the cover and concealment provided by woods and other close terrain. Firing positions that are ideal for defending against mechanized units maybe useless for defending against infantry units.

3. **Observation and Fields of Fire:** Seek to fire on the enemy when he cannot fire on you and do not let the enemy fire on you when you cannot fire on him.

4. **Plan of Defense/Scheme of Maneuver:** How the battalion commander plans to use the companies will influence the selection of firing positions. For example, if the battalion commander plans to surprise the enemy, firing positions should be selected that allow the fire of all or most of the weapons can be brought to bear on the enemy simultaneously.

The following paragraphs will discuss what forms a basic defensive firing position.

SECTOR OF FIRE

A sector of fire (fig. 4-3) is an area of responsibility assigned to a squad, a platoon, or a crew-sewed weapon to be covered by fire. The squad sector of fire is divided into fire teams and individual sectors of fire to cover the entire squad sector by fire. Sectors of fire ensure mutual support by overlapping the individual and fire team sectors. Normally, the squad leader is not assigned an individual sector of fire since the primary duty in the defense is directing and controlling the squad. Located within each sector of fire are lateral and forward limits.

1. **Lateral Limits:** Readily identifiable terrain features are selected to show the line of sight along each side of the sector. Two stakes are firmly placed near the position of a weapon to show the lateral limits for periods of reduced visibility.

2. **Forward Limits:** The forward limit is established at the range at which the weapon will open fire. When possible, a terrain feature is selected to locate the forward limit. This allows the squad leader a method of control to open fire on the enemy at a precise time to maximize effectiveness.

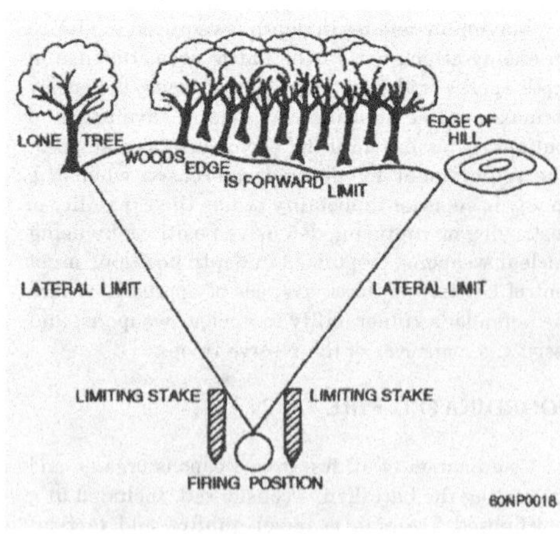

Figure 4-3.—Sector of fire.

FIGHTING POSITIONS

A fighting position is a location on the ground from which fire is delivered by an individual, a fire unit (squad or fire team), or a crew-served weapon. Before you select a fighting position, the assigned sector of fire is carefully examined from various locations. Examining is done in the prone position to ensure effective coverage of the sector of fire. The positions must provide the following:

- Desired support of the sector
- Good fields of fire
- Maximum use of available cover and concealment
- Facilitate effective fire control by the unit leader

Carefully examine assigned sectors of fire from the **prone** position, and then select a fighting position to ensure effective coverage of the sector. Primary, alternate, supplementary, and battle fighting positions must be established (fig. 4-4).

NOTE: Range cards must be completed for each fighting position. Range cards will be discussed further in this chapter.

Primary Fighting Position

The primary fighting position is the best available position from which the assigned sector of fire can cover. Individuals, fire teams, squads, and crew-served weapons are assigned primary positions.

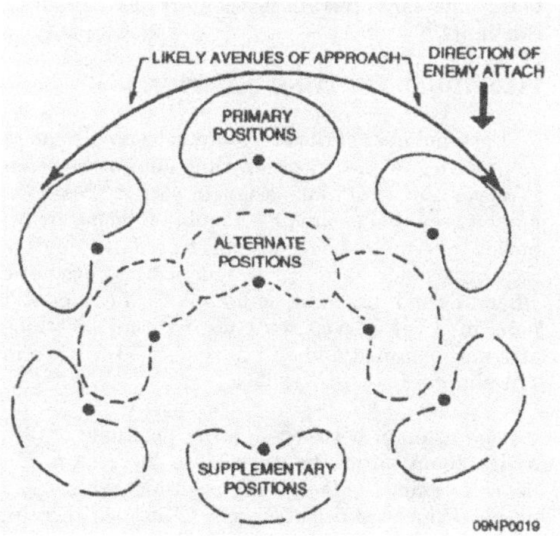

Figure 4-4.—Defensive firing positions.

Alternate Fighting Position

Alternate positions are not normally assigned to individuals or squads within the platoon. They are used primarily by crew-served weapons. An alternate position is located so that a crew-served weapon can continue to cover the original assigned sector of fire when the primary position becomes unsuited for carrying out that mission.

Supplementary Fighting Position

Supplementary positions are prepared to guard against attack from directions other than those from which the main attack is expected. A supplementary position is a secondary position that does not cover the same sector of fire as the primary position. Supplementary positions are for security reasons and ensure protection, when occupied against surprise enemy attack from an unexpected direction. Movements to supplementary positions and concealed routes are covered to avoid enemy detection.

Battle Position

The battle position is where the main effort of defense is concentrated. Companies and platoons are assigned battle positions. The battle position is made up of a series of sectors of fire that support and interlock one another. Based on the battle position of the company, platoon battle positions are assigned a right and left limit of fire in which gunfire can be delivered.

PRINCIPAL DIRECTION Of FIRE (PDF)

A principal direction of fire (PDF) is a specific direction within a sector of fire of a flat trajectory weapon including crew-seined weapons. The PDF is designated as the primary fire mission for that weapon. **Never assign more than one PDF per weapon.** Assign a PDF using a terrain feature. Using a stake to guide the PDF during reduced visibility is vital to the effective firing of a PDF (fig. 4-5).

The PDF is used as follows:

1. To cover a gap in a final protective line (FPL) of a crew-served weapon

2. To cover a specific terrain feature endangering the company or platoon battle position, such as draws,

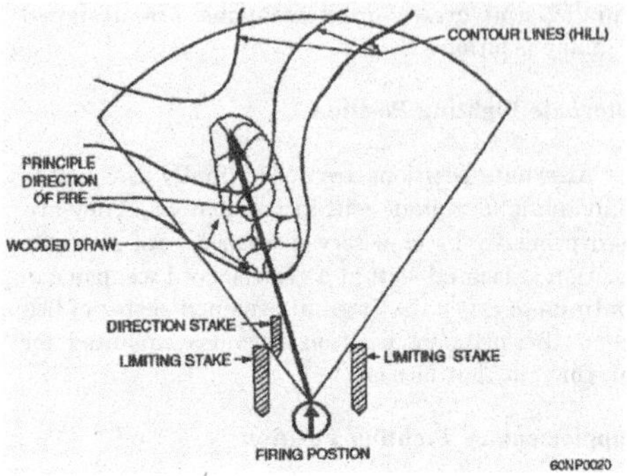

Figure 4-5.—Principal direction of fire.

that may seine as an avenue of approach or a hilltop that may serve as a possible enemy vantage point

3. To protect a crew-served weapon by firing across its front

4. To augment the band of flank fire placed immediately in front of the FEBA or the defensive line when them are no visible targets of opportunity to the front

FIRE TEAM IN THE DEFENSE

A primary duty of a squad leader is to organize the fire teams in the defense. This is accomplished by specifying a sector of fire for the squad and PDF for each automatic rifleman. The squad leader must inform the fire team leader of any crew-served weapons, augmented from the weapons platoon, which will help in covering the assigned sector of fire. The squad leader and the fire team leader will select terrain features to show lateral and forward limits of the assigned sector of fire. The mission of the fire team is to stop the enemy forward of the FEBA or to repel the enemy by close combat if the enemy reaches the FEBA.

FIRE PLAN

The fire team leader formulates a fire plan to cover the entire sector assigned by the squad leader with the heaviest possible volume of fire (fig. 4-6 and appendix II).

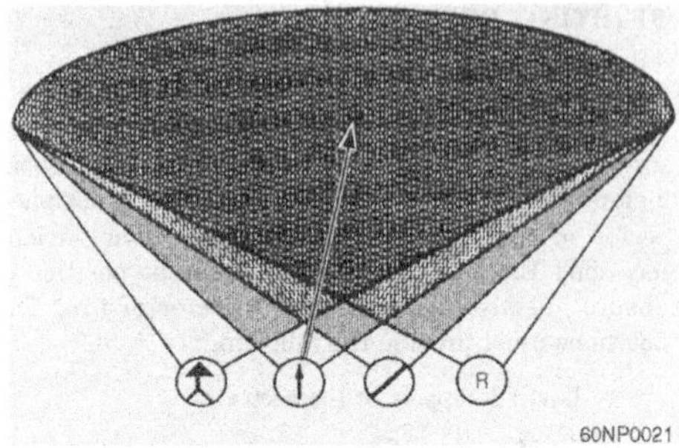

Figure 4.6.—Fire team fire plan.

The fire plan includes the following:

- Individual sectors of fire
- Individual fighting positions
- Automatic riflemen PDF assigned by the squad leader or platoon commander
- Crew-served weapon positions
- Key terrain features
- Position of the fire team leader

INDIVIDUAL SECTORS OF FIRE

If practical, each member in the fire team should cover the entire sector of fire of the fire team. The same terrain featues are used to show the limits.

INDIVIDUAL FIGHTING POSITIONS

Each member of the fire team is located so as to occupy an assigned position that allows the team to cover, by fire, the assigned sector. Positions maybe prepared as single-or double-fighting holes; however, double holes are preferred. If double-fighting holes are prepared, the automatic rifleman and assistant automatic rifleman will pair off. Crew-served weapons fighting positions are more detailed and will be discussed further in this chapter.

The interval between fighting positions within a fire team varies between 5 to 20 meters. In dense terrain, single-fighting positions are usually prepared and are 5 meters apart, In open terrain, single holes may be 10 meters apart; double-fighting positions, 20 meters apart.

AUTOMATIC RIFLEMAN

Since the automatic rifles are the backbone of the defense of the squad, the squad leader selects the exact fighting position. The remainder of the fire team is then positioned around the automatic rifles. The PDF of the automatic rifles are selected by the platoon commander or the squad leader.

RIFLEMAN

A rifleman is positioned so he can cover the entire fire team sector. The position must provide support and protection for the automatic rifleman.

GRENADIER

The squad leader positions the grenadier where the M203 grenade launcher is most effective. In combat, the grenadier is usually close to the squad leader. The squad leader may choose to use the grenadier to cover dead space in the defense.

FIRE TEAM LEADER

Usually the position of the fire team leader is at the center of the fire team from which he can perform the following:

- Observe the entire fire team and its sector of fire.
- Direct the fire of the automatic rifle.
- Observe the squad leader.

FIRE PLAN SKETCH

A sketch of the fire plan is submitted by the fire team leader to the squad leader. The magnetic North line provides a reference to show the direction the fire team is facing. Figure 4-7 shows the meaning of the various symbols.

SQUAD IN THE DEFENSE

The platoon commander assigns each squad a defensive mission specifying a sector of fire and a primary position. Terrain features are selected to show the lateral and forward limits of the sector of fire of the squad. The platoon commander designates the general firing positions and the PDF for specific automatic rifles or crew-served weapons, which are critical to the defense of the entire platoon. Designated on-call targets are established and coordinated with the Combat Operation Center (COC). Supplementary positions are assigned to the squads for all-around defense and to protect the flanks or the rear of the platoon (fig. 4-8 and appendix II).

SQUAD FIRE PLAN

The squad leader formulates a basic squad fire plan to occupy physically the assigned primary position and to cover, by fire, the sector assigned by the platoon commander. The fire plan includes the

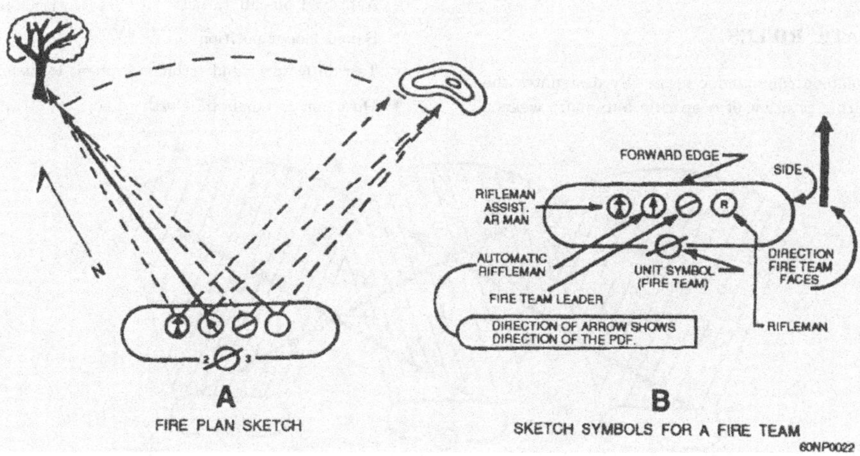

Figure 4-7.—Fire plan sketch and symbols.

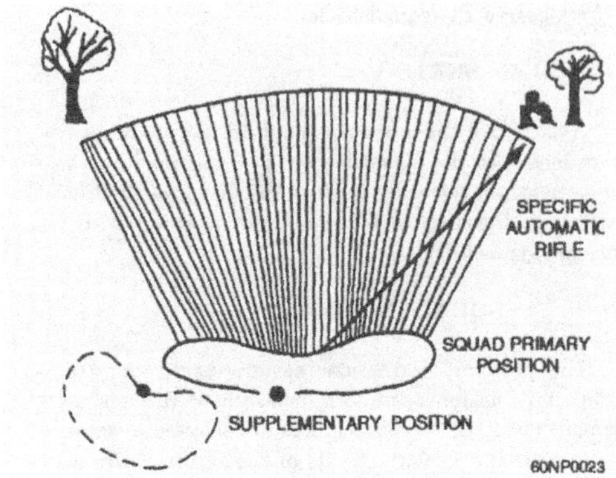

Figure 4-8.—Squad position.

assignment of the sectors of fire of the fire team and PDF for automatic rifles or crew-served weapons (fig. 4-9).

FIRE TEAM POSITIONS

Fire teams are distributed so they physically occupy the assigned position and can cover, by fire, the assigned sector. The fire teams are normally placed abreast, firing the FEBA. Terrain dictates the effective placement of the individual members of each fire team. Place crew-served weapons first, so the position of the squad can provide close-in protection for these weapons.

AUTOMATIC RIFLES

The platoon commander norm ally designates the general firing position of a specific automatic weapon or crew-served weapon; however, this responsibility y may be delegated to the squad leader. If so, the squad leader will designate the PDF for each automatic weapon or crew-served weapon.

SQUAD LEADER POSITION

The position of the squad leader is usually slightly to the rear of the fire teams and to the center of the squad. From this position the squad leader should be able to observe the following:

- Assigned sector of fire.
- Observe as much of the squad as possible.
- Maintain contact with the platoon commander.

SQUAD FIRE PLAN SKETCH

Based upon the fire plan sketch received from the fire team leaders, the squad leader prepares a squad fire plan sketch. Two copies are prepared-one for the squad leader and the other for the platoon commander (fig. 4-10 and appendix II).

The sketch should include the following:

- Fire team positions and sectors of fire
- Position and PDF of the automatic rifles
- Crew-served weapons position
- Primary fire missions (FPL or PDF)
- Approved on-call targets
- Squad leader position
- Terrain features and estimated ranges to them
- Direction of magnetic North

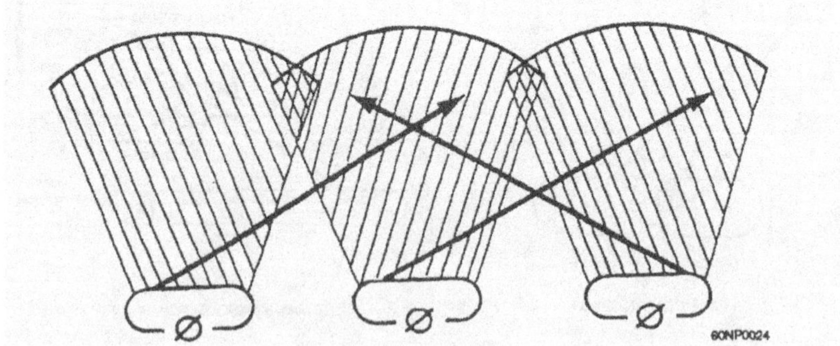

Figure 4-9.—Squad fire plan.

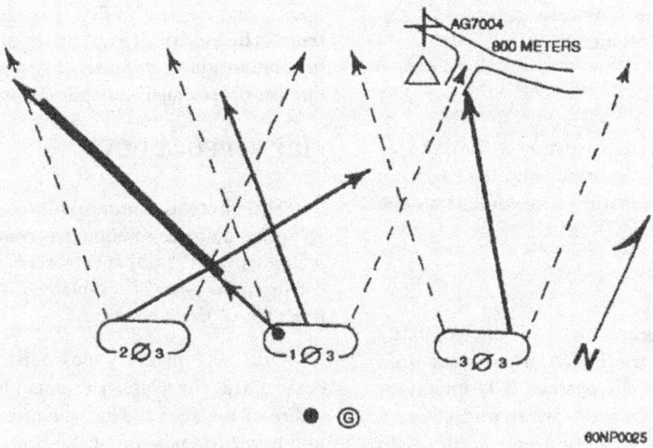

Figure 4-10.—Squad fire plan sketch.

PLATOON IN THE DEFENSE

Located within the assigned company battle area is the platoon defense area. The company commander assigns the platoon a frontage. The frontage left and right limits are taken from a reference point on the terrain or the flank of an adjacent platoon. A distance from the FEBA to the rear is also described (fig. 4-11).

FRONTAGE

The terrain and the size of the company battle area dictates the frontage assigned to a platoon. Physical

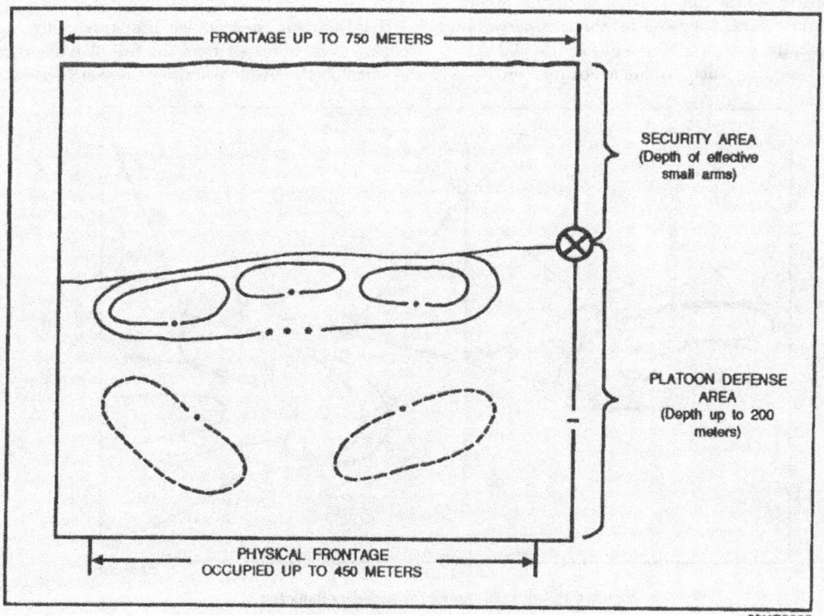

Figure 4-11—Frontline platoon defense area.

frontage is the area that is physically occupied by the platoon. It can be up to 450 meters.

DEPTH

The depth of a platoon defense area is the distance between the squad primary positions and the rearward extension of their supplementary positions. It maybe up to 200 meters.

SECURITY AREA

The platoon is assigned part of the company security area forward of the FEBA. This is the area where the company sends out patrols. It is important as a platoon commander to know when and where a patrol will cross into his security area Coordination between platoon commanders and the company commander on the positions of listening posts, observation posts, and early warning devices is essential.

PLAN OF DEFENSE

The plan of defense of the platoon commander consists of a well-developed fire support plan that is integrated with the fire support and barrier planning of the company. Close combination with the other platoon commanders assigned to the company is critical. Basically, the plan of defense of the platoon consists of assigning each squad a position on the terrain and a sector of fire. The plan of defense results from the estimate of the situation and the fundamentals of defense of the platoon commander discussed previously in this chapter.

FIRE SUPPORT PLAN

The platoon commander coordinates with the weapons platoon commander concerning crew-served weapons and final protective lines (FPLs). The defensive fire plan of the platoon is integrated with the fire plan of the company.

Once the primary positions of the squads are determined, the platoon commander selects the sector of fire of the squad. The sectors of the squads overlap and cover the portion of the company battle position assigned to the platoon. The sectors of fire of the flank squads overlap the adjacent platoon's flank squad's sectors of fire, and the interlocking fire provides mutual support (fig. 4-1 2).

When a gap exits between adjacent platoons, proper coverage is accomplished by use of indirect fire (fig. 4-13).

PLATOON FIRE PLAN SKETCH

The platoon commander prepares a fire plan sketch or overlay based from the fire plan sketches of the squad leader. A platoon fire plan sketch or overlay

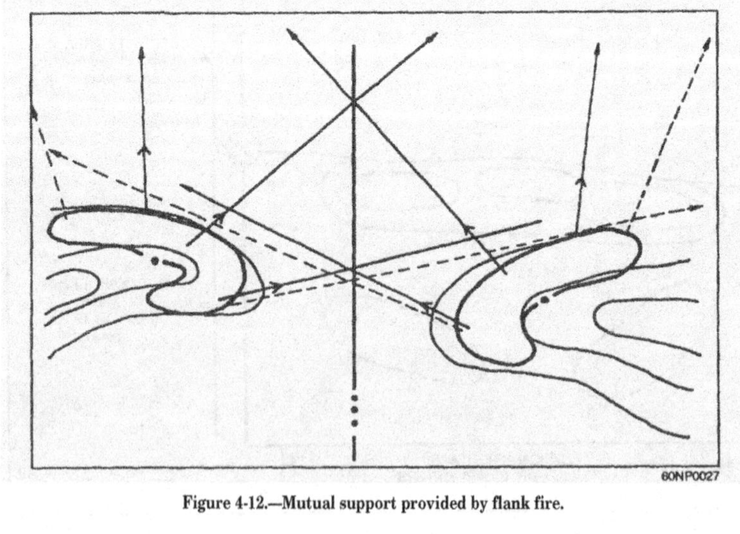

Figure 4-12.—Mutual support provided by flank fire.

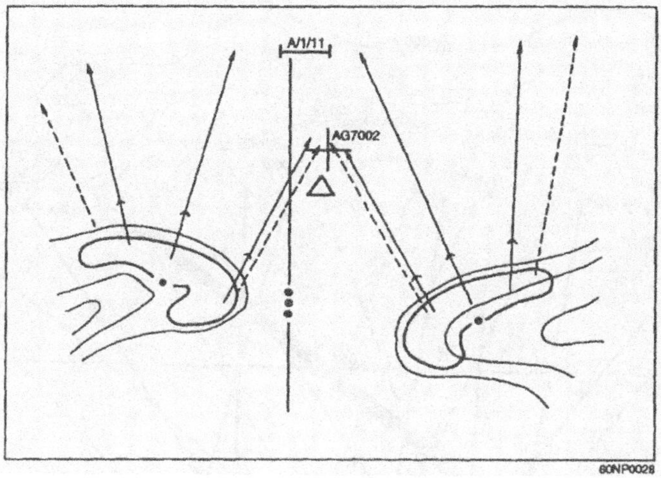

Figure 4-13.—Coverage for a gap.

is submitted to the company commander for approval (fig. 4-14 and appendix II).

The sketch or overlay includes the following:

- Squad primary position and sectors of fire
- Positions and PDFs for all automatic rifles including crew-served weapons
- Location of platoon or company observation post
- FPLs of all weapons located in platoon defense area
- On-call targets
- Barriers and early warning devices
- Claymore mines

ORGANIZING THE PLATOON IN THE DEFENSE

The task of organizing the platoon in the defense begins immediately upon arrival of the company. Work commences in the order of priority established by the company commander. Top priorities for the platoon normally include the following actions in order of importance:

- Establish security.
- Hastily set up a communication network.
- Position automatic and crew-served weapons.
- Clear fields of fire.
- Assign sectors of fire and PDFs.
- Prepare fighting positions.
- Plan, coordinate, and plot available fire support.
- Install tactical and supplementary wire.
- Lay and bury wire for the final communication network.
- Prepare other obstacles including claymores and protective wire.
- Prepare alternate positions.
- Prepare supplementary positions.

NOTE: The platoon commander must have constant communication with the squad leaders and the company CP.

CREW-SERVED WEAPONS

Crew-served weapons are the fire support elements of the rifle companies. The weapons platoons and the 60-mm mortar platoons are responsible for the crew-served weapons. Their purpose is to provide each company with organic machine gun support, mortar fire support, and antitank defense capability.

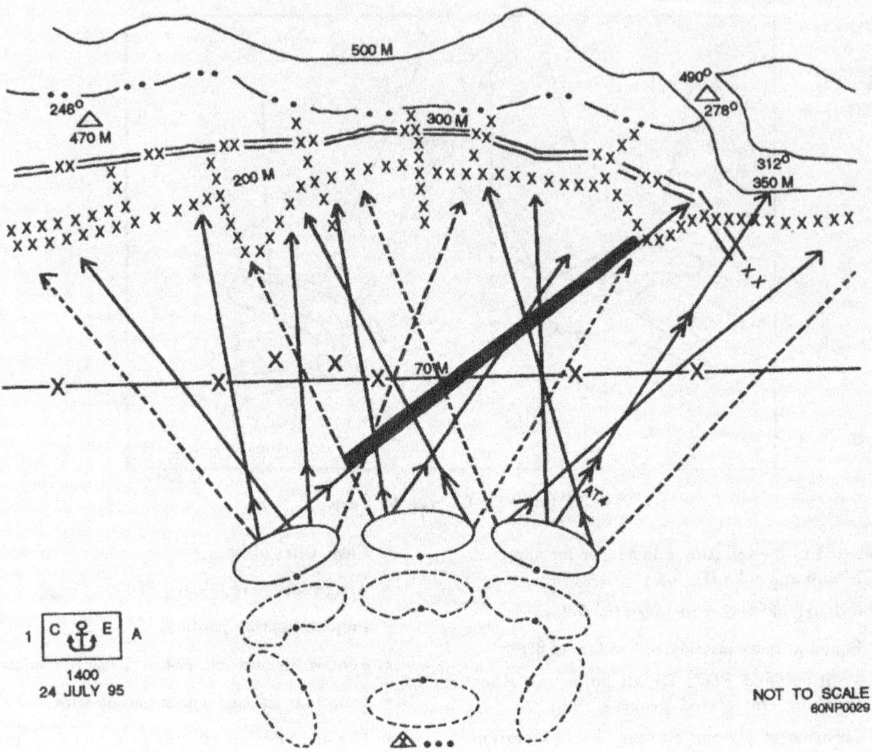

Figure 4-14.—Platoon fire plan sketch.

WEAPONS PLATOON

Alfa, Bravo, Charlie, and Delta companies each have a weapons platoon. Terrain will dictate the number of crew-served weapons assigned to each platoon. The weapons platoon is normally composed of the following crew-sewed weapons:

- M60E3 machine gun
- .50 caliber machine gun
- M19 grenade launcher
- M136, 84-mm launcher AT4 antitank weapon

NOTE: Refer to the *Seabee Combat Handbook,* Volume 1, for the characteristics of crew-served

Weapons Platoon Commander

Similar to the rifle platoons, the weapons platoon commander must have constant communication with the squad leaders responsible for the crew-served weapon teams. The weapons platoon commander conducts his reconnaissance with the company commander or is directed to conduct it separately. In either case, the weapons platoon commander plans the employment of the weapons platoon. During the planning, it is important that the weapons platoon commander works with the rifle platoon commanders. The planning includes the following:

- Fighting positions for crew-served weapons
- FPLs and PDFs for machine guns
- Primary and alternate positions for machine guns

- Sectors of fire for each assault team
- Covered waiting positions for the assault teams
- Primary and alternate positions for the assault teams

The selection of the weapons platoon commander is the substance of recommendations made to the company commander for the primary defensive employment of the weapons platoon. While on reconnaissance, the weapons platoon commander also selects supplementary positions, as necessary, from which the machine gun team may perform other defensive missions.

Fire Support Plan

The company commander exercises control of the weapons platoon through the weapons platoon commander. He exercises control for the most effective defensive employment of the company. The weapons platoon commander ensures its effectiveness by preparing a detailed plan consistent with the recommendation accepted by the company commander. A fire support plan is prepared by the weapons platoon commander. The plan is based upon an estimate of the situation, the fundamentals of defense previously discussed, and a sound knowledge of crew-sewed weapon employment.

M60E3 Machine Gun Section

The machine gun section is a major element of fire support for stopping the enemy assault immediately in front of the battle area. In a well-organized defense, most of the final protective fire (FPF) of the company is delivered by the machine gun section. The positions of the machine guns are planned primarily for this purpose. Roles of the M60E3 machine gun section are as follows:

- Support the rifleman in the offense and defense.
- Provide heavy volumes of controlled, accurate, and continuous fire.
- Engage predetermine targets under all visibility conditions.
- Provide long range, close defensive, and final protective fries (FPFs).
- Basis for FPFs in the overall defensive fire plan.
- Cover likely avenues of approach.
- Provide grazing fire.

Use of .50 Caliber Machine Gun

In general, the roles of the .50 caliber machine gun are similar to the M60E3. The major difference is that the .50 caliber machine gun ammunition can penetrate light armor. The additional roles of the .50 caliber machine gun are as follows:

- Provide protection for motorized movements.
- Destroy lightly armored vehicles.
- Defend against low-flying hostile aircraft.

M19 Grenade Launcher

The role of the M19 grenade launcher is similar to the M60E3 and .50 caliber machine guns. The weapon can be used to provide indirect or direct firing. Like most machine guns, the M19 grenade launcher can be mounted on a vehicle.

AT4 Antitank Weapon

Primarily, the M136, 84-mm AT4 antitank weapon is used against armored personnel carriers; however, it can be used to disable a battle tank by striking the side or rear of the tank The AT4 can be fired from the right shoulder only.

MACHINE GUN TACTICAL EMPLOYMENT

The basic unit of machine gun employment in defense is the squad. Machine gun squads consist of a squad leader and two four-man machine gun teams. Machine guns used on the **FEBA** are normally used by squads.

Employment of Machine Guns

Machine gun squads used on the FEBA are normally assigned a final protective line (FPL) and a sector of fire or a principal direction of fire (PDF). Both machine guns of the squad fire the same general FPL and sector of fire from positions a minimum of 35 meters apart.

Based on terrain, it may be necessary to split some squads to provide effective machine gun coverage. A machine gun squad is split when each of its guns has been assigned a different firing mission; that is, a different final protective line or principal direction of fire and sector of fire. A machine gun squad is split only when necessary. If the squad is split, the two guns should be used as close to each other as the machine gun fire plan will allow. This action ensures interlocking fire, ease of control, and supplies (fig. 4-15).

Machine Gun Final Protective Lines

Effective final protective fire is characterized by the following:

1. **Flanking:** Maximum flanking fire is desirable. The more frontal the fire, the less effective the coverage of the company front.

2. **Interlocking:** Interlocking fire adds to the effectiveness of the fire plan. Such fire reduces the number of gaps in the final protective lines and provides mutual support between adjacent units.

3. **Grazing:** Final protective lines are located to obtain maximum grazing fire. Grazing fire is fire in which the trajectory of the bullets does not rise above the height of a man, standing. On flat or uniformly sloping terrain, machine gun fire grazes to a maximum range of 700 meters from the gun. Figure 4-16 shows a proper technique for graphically displaying final protective fire and gaps in its grazing fire on an overlay or sketch.

Machine Gun Sector of Fire

A sector of fire is assigned to each machine gun squad. A machine gun sector of fire does not normally exceed 800 roils (45 degrees). Adjacent machine gun squad sectors should overlap. Preferably, the final protective line comprises the near boundary of the sector. It maybe located within the sector when the grazing fire is slightly more frontal than desirable and machine gun fire coverage closer to the FEBA is required.

Machine Gun Principal Direction of Fire

When the terrain does not allow for an effective final protective line, machine guns on the FEBA may be assigned principal directions of fire for covering dangerous avenues of approach. In such situations, the principal direction of fire may fall within the sector of fire or comprise either of its boundaries.

NOTE: A machine gun cannot be assigned a final protective line (FPL) and a principal direction of fire (PDF).

Machine Gun Communications

Whenever possible, the primary positions of the machine gun squad are provided with sound-powered telephone communications on the company wire net

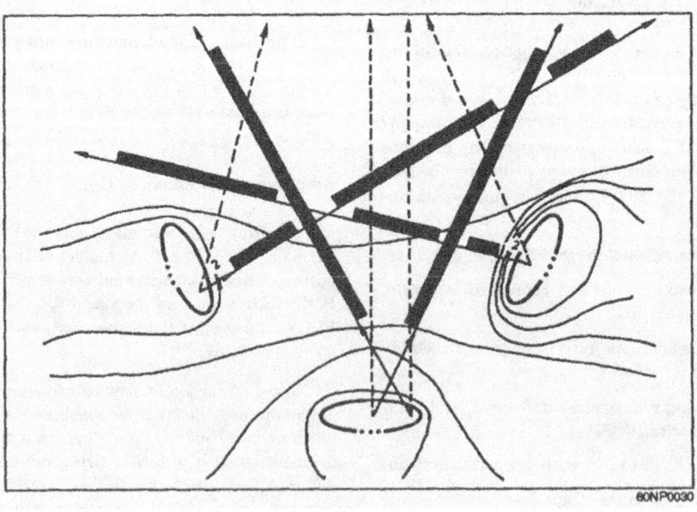

Figure 4-15.—Machine guns split.

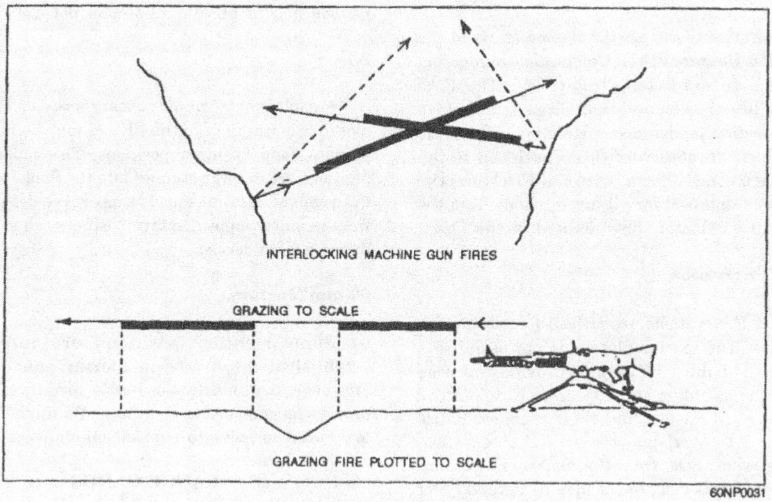

Figure 4-16.—Machine guns FPL and grazing fire.

to the weapons platoon commander. If time permits, wire is laid to alternate and supplementary positions. Messengers are used when wire communications are inoperative or have not been installed.

AT4 TACTICAL EMPLOYMENT

The assault squad is the basic unit for the tactical employment of the AT4 antitank weapon. Because of the short range of the AT4, they are normally positioned with the frontline platoons to provide close-in antitank defense. The weapons platoon commander normally uses a messenger to communicate with the assault squads.

Positions of Assault Squads

The location of the best observation and fields of fire covering the avenues of approach for armor vehicles dictates the positioning of assault squads. A primary and several alternate firing positions are prepared from which each avenue of approach can be covered. When the armor threat develops, the squad moves by covered routes to the previously prepared positions.

AT4 Sectors of Fire

Each assault squad is assigned sectors of fire to ensure that all avenues of approach for armor vehicles are covered and to provide overlapping areas of antimechanized responsibility. The size of the sector is limited only by the available observation and fields of fire. The assignment of a sector of fire does not preclude firing at targets outside the sector. When the squad is used as a unit, both assault teams are assigned the same sector. When the squad is split, the two teams may be physically located in proximity but are responsible for separate sectors. Fire planning should avoid splitting assault squads unless absolutely necessary.

60-MM MORTAR TACTICAL EMPLOYMENT

In the defense, final protective fire of a single mortar covers an area of approximately 50 by 50 meters. A FPF may be located within 60 meters of the FEBA. Thus the 60-mm mortar can cover small gaps or dead space in the machine gun FPF close to the FEBA. Firing positions for mortars should be located between 150 and 300 meters to the rear of the FEBA.

Communications

Communication to the mortar platoon starts at the Fire Direction Center (FDC). Companies requesting a fire mission do so by contacting (FDC). The FDC will plot the fire mission and issue fire commands to the mortar section leader responsible for the mortar gun crew. Forward observers (FOs) assigned to the mortar platoon communicate with the FDC directly. FDC must seek approval for all fire missions from the Fire Support Coordinator (FSC) located in the COC.

Sectors of Observation

Sectors of observations are critical for mortar fire to be effective. The mortar platoon is not assigned a sector of fire but must be able to provide coverage throughout the battalion security area. Rifle platoon commanders must ensure that their assigned battle area is covered by overlapping sectors of observation. Forward observers (FOs) from the mortar platoon are usually assigned this task, but it may be necessary for the rifle companies to assist the FOs. Personnel from the rifle companies assigned as FOs must be able to call in a fire mission and do so via the platoon commander and the company CP.

Rate and Duration of Fire

In the absence of any instructions, the normal rate of fire is nine rounds per minute for the first 2 minutes and six rounds per minute after that for the duration of final protective fire (FPF). The FDC chief will normally specify the number of rounds to be expended on each target to the section leader. The COC, upon approval from the battalion commander, sets the number of rounds to be fired for FPF. Consideration is given to the ammunition on hand, nonorganic support available, anticipated resupply schedule, area to be covered, and enemy activity.

SUPPLEMENTARY MISSIONS

Both the machine gun squad and the assault squad may be positioned and employed in performing supplementary missions in defense.

Machine Guns

Machine gun squads, employed on the FEBA or in the reserve area, prepare supplementary positions from which to cover sectors of fire **different from their primary sectors.** Missions are usually based upon an enemy penetration of an adjaent company or platoon and to provide all-around defense.

AT4

Assault squads, whose primary missions are to cover avenues of apprach to the FEBA, may be required to prepare supplementary positions. The supplementary positions are to cover avenues into the flanks and rear of the company. Assault squads support the defense by fire from planned supplementary positions when an armor threat fails to develop.

60-mm Mortars

Supplementary positions are infrequently established for a 60-mm mortar squad section. However, fire within the battle area is planned to provide for supporting frontline rifle units when they are forced to move to supplementary positions.

RANGE CARDS

Once a machine gun team is deployed in a defensive position and the gun laid, the squad leader must ensure that range cards have been completed. Two sets of range cards must be made for every fighting position. Range cards are very important and are used for several purposes. The squad leader forwards one copy as soon as possible to the platoon commander. The platoon commander ensures the card is accurate and then forwards it on to the company CP. Range cards are the basis for the fire plan of the company. All supporting fire, security sectors, and defensive coordination are planned using machine gun fire as the base from which to build.

The second copy of the range card will be kept by the gun team. The card is used by members of the gun team to fire at predetermined targets or to help in range estimations to other targets. Range cards may be passed on to a relieving battalion. Patrols from companies or other units may want to look at the range cards to identify final protective lines. Patrols need to know of any dead space to ensure that they are clear of infiltrators and where to take cover during an FPF.

COMPLETING A RANGE CARD

Range cards must be neat, clear, and prepared using a universal format. Observe the range card symbols and sample range card shown in figures 4-17 and 4-18 as each component is explained below. (see appendix II)

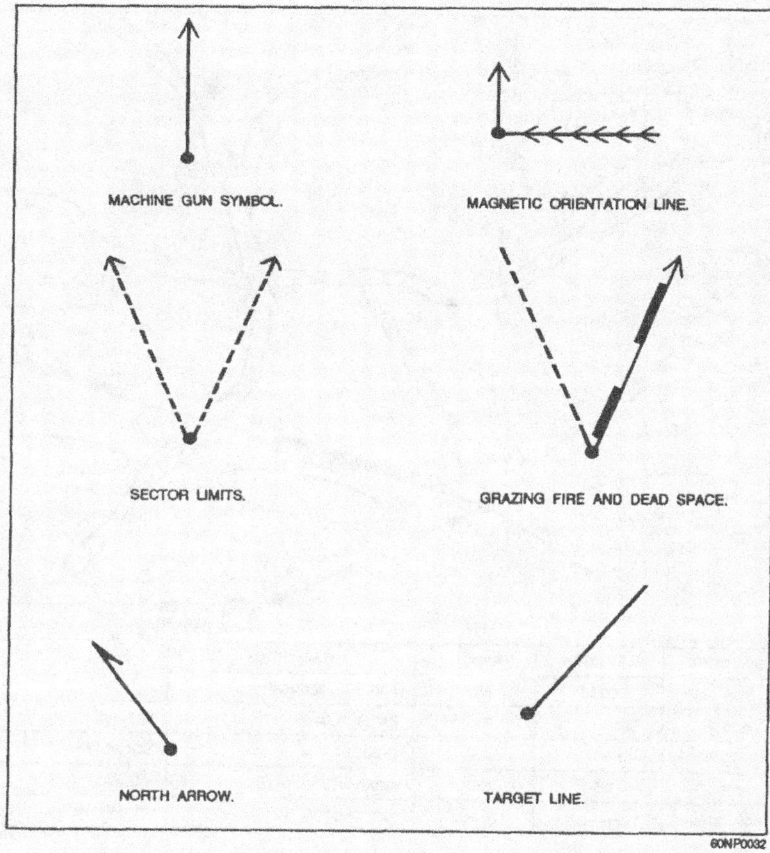

Figure 4-17.—Range card symbols.

1. **Identification Block:** This block identifies what gun the card was drawn for and on what date it was drawn. For security reasons, the unit should only be identified to company level.

2. **Weapon Symbol:** The correct symbol for a machine gun is a single dot with a solid arrow extending out in the direction of the FPL or PDF. The dot represents the location of the gun. Record the magnetic azimuth of the FPL or PDF along this line.

3. **Magnetic North Arrow:** This arrow is drawn from the dot that represents the gun position in the direction that represents North from the gun position if you were looking out toward your sector of fire. The correct symbol for North is au arrow with a single barb.

4. **Magnetic Orientation Line and Location Grid Coordinates:** These are the two methods used so that the CP, the battalion, or other units can positively locate a gun position. The magnetic orientation line method is a line drawn from a prominent terrain feature that is located preferably behind friendly lines. **It is a single-solid line drawn from a prominent terrain feature to the gun position with several arrow heads pointing in the direction of the gun position; the magnetic azimuth in roils from the prominent terrain feature to the gun position is recorded on the line.**

NOTE: One degree is equal to 17.7778 mils. Example: To convert an azimuth of 140 degrees,

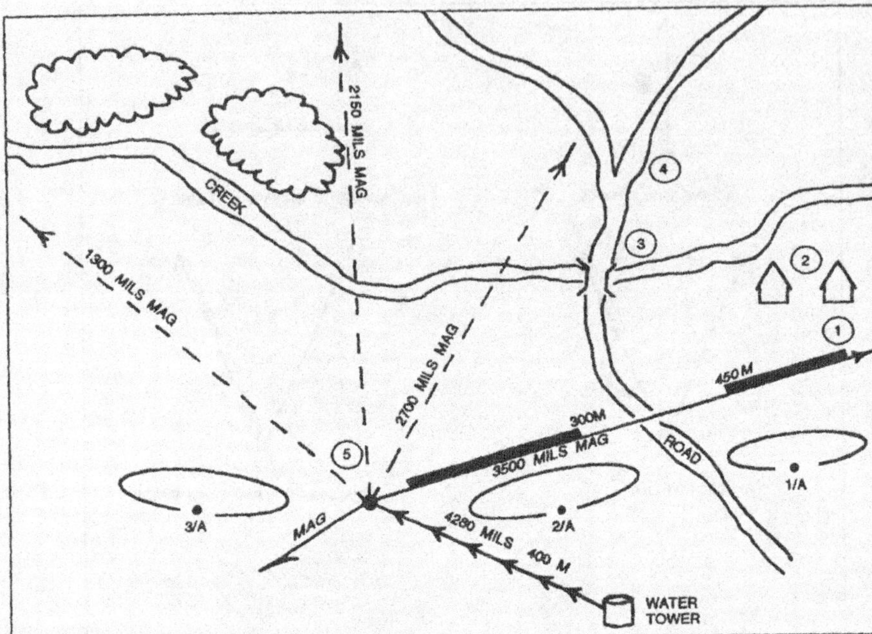

NO.	DIRECTION	ELEVATION	RANGE	DESCRIPTION	
1	3500 M/L	+50/25	650	FPF D.S. 300-450	
2	200 R	-50/10	600	FARM BLD	GUN #1
3	95 L	- 0/37	400	BRIDGE	SOUND #3
4	290 L	- 0/5	550	ROAD JCT	CO. "A"
5	1300-2150	——	500-900	ALT SECTOR	18 JAN 89

80NP0033

Figure 4-18.—Range card with an FPL.

simply multiply 140 by 17.7778 which equals 2488.892. Round up to 2,489 mils.

The grid coordinates method is where the grid of the gun position is recorded next to the dot in the machine gun symbol. Only one of these two methods to locate the position of the gun is needed.

5. **Sector Limits:** Sector limits are drawn as broken lines ending in arrow heads. When you are using an FPL, only one sector limit will be drawn because one is located in the same location as the FPL.

6. **Grazing Fire:** If an FPL is used, a heavy-shaded area is drawn along the inside of the FPL, which will show the limits of the grazing fire. Show any dead space by breaks in the grazing fire by shaded areas. Record the near and far limits of the dead space in meters or record the range next to the ends of the shaded areas.

7. **Terrain Features:** Draw only those terrain features that significantly add to the clarity of the range card. If a terrain feature is drawn, draw it to the correct perspective.

8. **Location of Friendly Troops/Equipment:** Draw in any friendly positions or equipment that is either in or near the sector limits of the machine gun. Clearly label them with both a description, range, and direction.

9. **Targets:** Draw targets to perspective and label them with a number. The number one target will be

either the FPL or the left sector limit. If the FPL is on the right sector limit, number all other targets sequentially from right to left; at all other times, number targets from left to right. There are two approved methods for recording target data. One method is to record data directly on the sketch along the line leading to the target. Another method is to use a data block at the bottom or reverse side of the range card.

Range Card for an FPL

The following scenario is to illustrate how a range card is prepared for an FPL. Figure 4-19 provides an illustration of a tactical field setting to help you.

Scenario

A machine gun fire team from the weapons platoon is attached to the second platoon of Charlie Company deployed along a low ridge overlooking a narrow valley. The weapons platoon squad leader, along with the other gun team, is supporting another company. The creek bottom across the valley is suspected of being a major infiltration route for the enemy. On the second platoon's right limit is the first platoon. Bravo Company joins the second platoon on the left limit. The second platoon commander has informed the machine gun team leader that the gun is to be positioned on the left flank of the platoon. The team leader informs the machine gun team to lay the gun to be able to fire an FPF across the front of the ridge. The line formed by the base of the ridge is to be the right sector limit. The trees at the bend in the creek in front of the gun position is the left limit. Interlocking fire for the FPF will be obtained from the third squad gun team attached to the first platoon. Using a compass or GPS, the gun team has located the gun at grid coordinate 94576259. The grid is also 750 meters from the water tower that is located on a magnetic back azimuth of 5,980 mils. The FPL lies on a magnetic azimuth of 4,250 mils.

Figure 4-20 shows the beginning stages of the range card. It shows the weapon symbol, the magnetic North arrow, the magnetic orientation line and location grid coordinates, the sector limits, and the grazing fire.

Figure 4-21 shows the completed range card. Details have been added, such as the unit identification block, the terrain features, the location of friendly troops or equipment, and the targets with their individual data.

Range Card for a PDF

The range card for a PDF is very similar to an FPL range card. The only real difference is the machine gun symbol. A PDF does not align with a sector limit.

The following scenario is to illustrate how to prepare a range card for a PDF. Figure 4-22 provides an illustration of a tactical field setting.

60NP0034

Figure 4-19.—Panoramic view of tactical area.

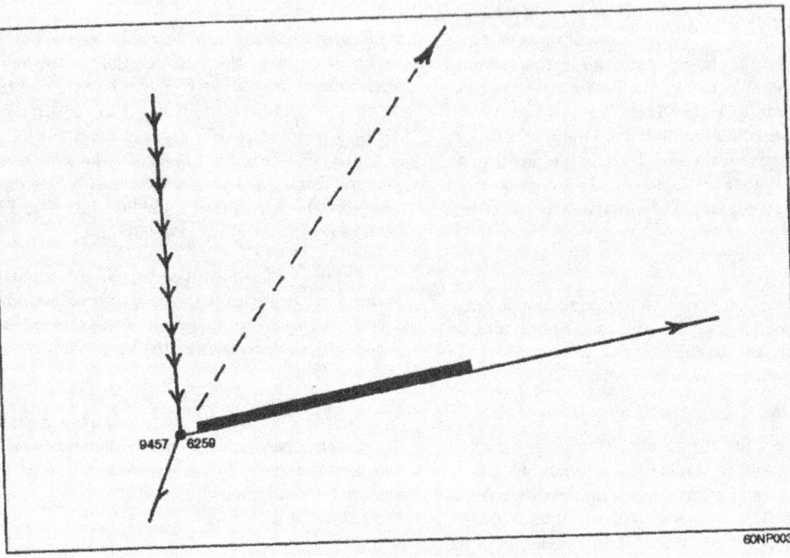

Figure 4-20.—Beginning of a range card.

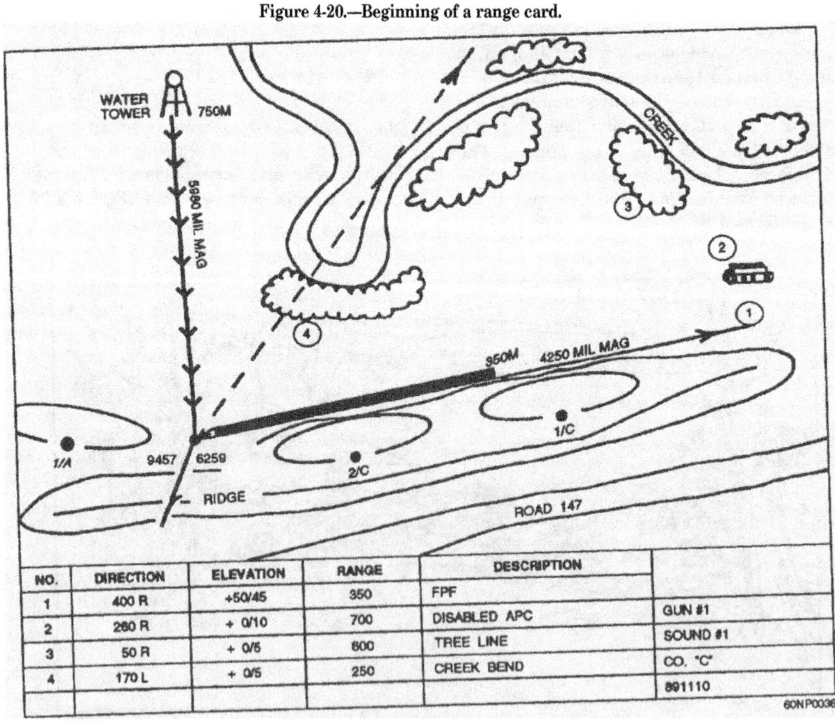

NO.	DIRECTION	ELEVATION	RANGE	DESCRIPTION	
1	400 R	+50/45	350	FPF	
2	280 R	+ 0/10	700	DISABLED APC	GUN #1
3	50 R	+ 0/5	600	TREE LINE	SOUND #1
4	170 L	+ 0/5	250	CREEK BEND	CO. "C"
					891110

60NP0036

Figure 4-21.—Completed range card.

Figure 4-22.—Panoramic view of tactical area.

Scenario

A machine gun fire team has been attached to the third platoon of Alfa Company. The squad leader and another gun team have become casualties. The platoon is deployed in a tree line overlooking a bridge. The enemy is suspected to have infiltrated the village on the other side of the river. The mission of the platoon is to stop any attempt from the enemy to cross the river via the bridge. The first platoon is to the right of the third platoon, and Bravo Company has joined the third platoon on the left. The commander of the third platoon has informed the machine gun team leader that the machine gun fire team is to be deployed on the left flank of the platoon. The team leader is to get the best angle to fire across the bridge. The team leader informs the machine gun fire team to lay the gun on a PDF centered on the bridge. The lone tree on the immediate right is the right sector limit, and the leftmost edge of the grove of trees across the river is the left sector limit. Using a compass or GPS, the machine gun fire team has located the gun at grid coordinate 468262. The grid is also 350 meters from a road junction that is located on a magnetic back azimuth of 5,420 mils. The PDF lies on a magnetic azimuth of 4,120 mils.

Figure 4-23 shows the beginning stages of the range card. It shows the weapon symbol, the magnetic North arrow, the magnetic orientation line and location grid coordinates, the sector limits, and the PDF.

Figure 4-24 shows the completed range card. Details have been added, such as the unit identification block, the terrain features, the location of friendly troops or equipment, and the targets with their individual data.

EXPEDIENT METHODS OF LAYING FOR PREDETERMINED FIRE

During different tactical situations, a T and E mechanism (refer to *Seabee Combat Handbook,* Volume 1, for predetermined fire) may not be available. This may be due to the lack of either a T and E mechanism or a tripod or a system is needed that is more applicable to firing at night. Whatever the reason, the following methods can be just as effective:

1. **Muzzle Stakes:** This method is good for night firing. It requires no illumination to lay on a target and can be used with or without a T and E mechanism. It is most effective when used with your gun mounted on a tripod. It can be used from a bipod mount if a notched stake is used to stabilize the rear of the weapon. Each target is designated by driving a stake into the ground under the muzzle. Drive the stake into a position and to a depth that will result in the gun being correctly aimed when the muzzle is lowered directly over the stake (fig. 4-25). The gun is tripod-mounted and will use swinging traverse fire.

2. **Notched Log under Muzzle:** Instead of using individual stakes, this method uses a log or board under the muzzle. This method is also used for night fire. Notches are cut into the log or board to the depth and position required to fire on each target. This method is very limited in depth, but if the ground that is fired over is level or uniformly sloping, this may not be a problem (fig. 4-26).

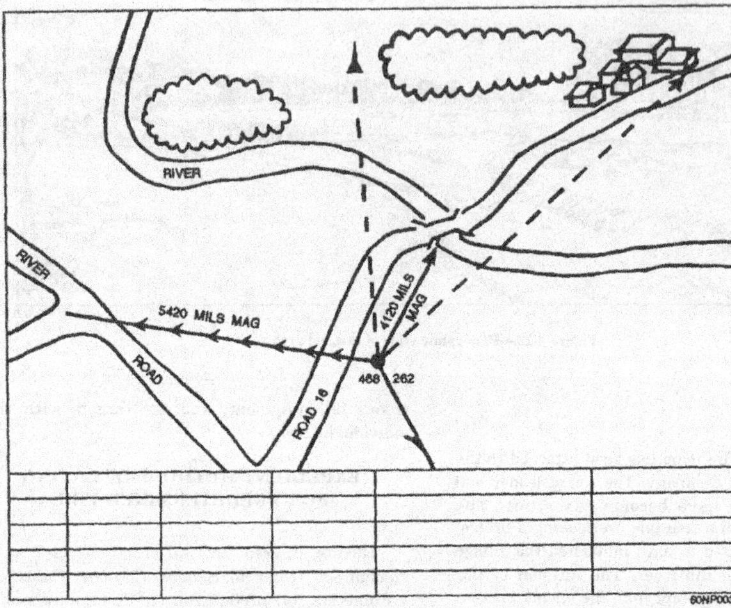

Figure 4.23.—Beginning of a range card.

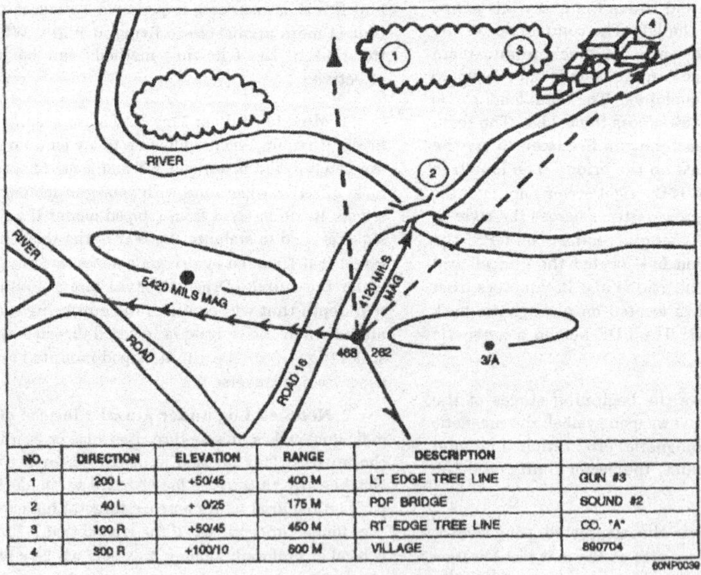

NO.	DIRECTION	ELEVATION	RANGE	DESCRIPTION	
1	200 L	+50/45	400 M	LT EDGE TREE LINE	GUN #3
2	40 L	+ 0/25	175 M	PDF BRIDGE	SOUND #2
3	100 R	+50/45	450 M	RT EDGE TREE LINE	CO. "A"
4	300 R	+100/10	800 M	VILLAGE	890704

60NP0039

Figure 4-24.—Completed range card.

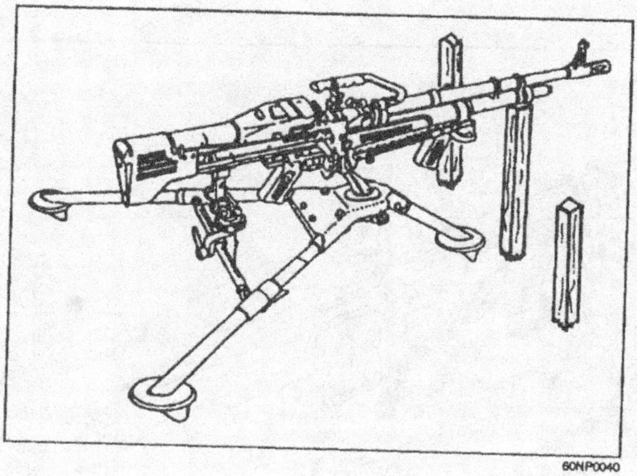

Figure 4-25.—Muzzle stakes.

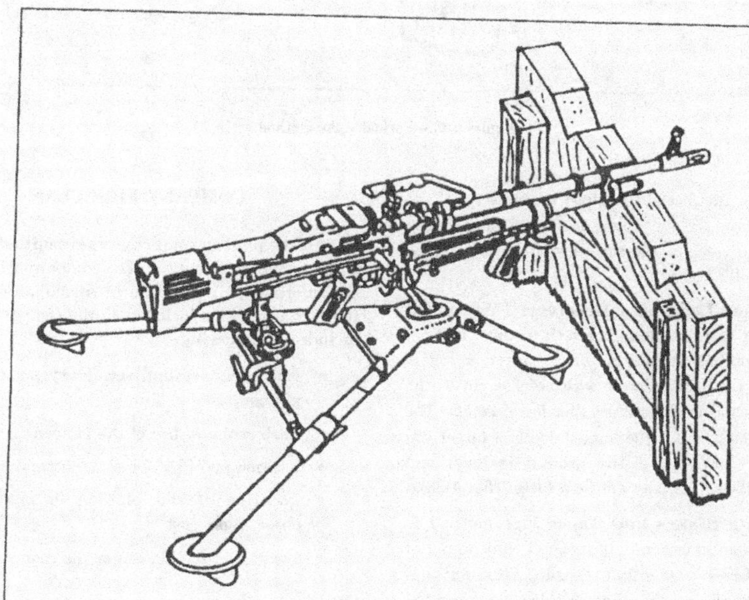

Figure 4-26.—Notched log method.

3. **Forked Stakes:** This is a method that works well with your gun supported by a biped only. To use this method, make a set of stakes with notches in the top or find naturally forked stakes. The front of the gun is supported by the biped, and the rear is supported by the forked stakes. It is essential that a set of U-shaped pits

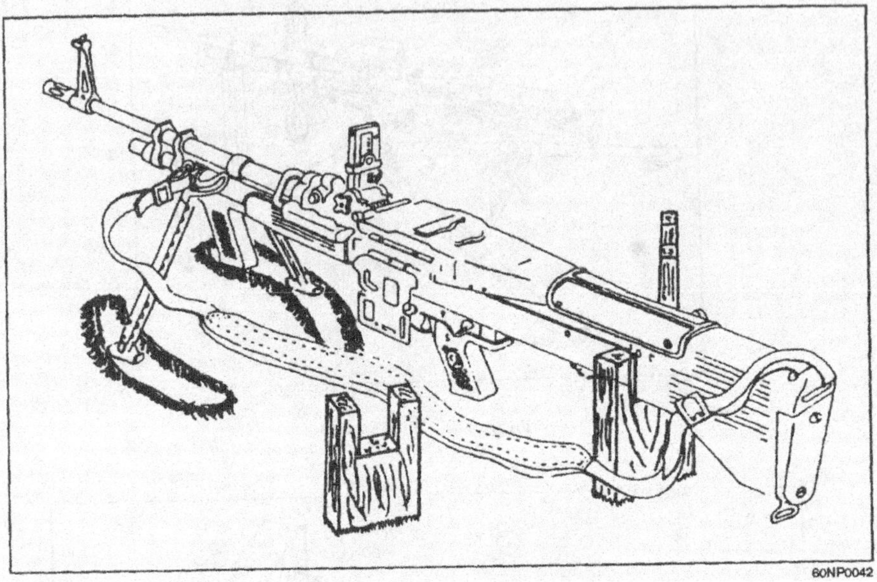

Figure 4-27.—Forked stake method.

are dug under the feet of the bipod to rotate in. If the feet of the bipod move about freely without being contained in these pits, the accuracy of the fire will be reduced (fig. 4-27).

4. **Notched Log under Receiver:** This method works similar in principle to the fork method. Instead of having notched stakes for each target, a log is substituted under the receiver with notches cut to the correct depth and in the correct spot for direction. The biggest drawback here is the limited depth of target you can engage; however, if the ground is level and uniformly sloped, this may not be a factor (fig. 4-28).

5. **Aiming Stakes and Tape:** This method is unique because it is the only method that does not restrict the free movement of the gun. Although it is more accurate than most others, it also consumes the most time and requires the use of some material that will not always be available. To use this method, you must mount the gun on a tripod and have the T and E mechanism in place. To implement this method, you will need luminous tape and one stake for each target and each sector. Figures 4-29 and 4-30 show this method with an M19.

COMPANY FIRE PLAN

All the platoon commanders submitted their fire plans to the company CP. The company CP compiles the information and prepares a company fire plan. Items that must be included on the company fire plan include the following:

- Each primary and secondary position of the platoon
- Each sector of fire of the platoon
- Position and PDF for each automatic rifle
- Location of the platoon or the company observation post
- Location of the platoon or the company listening post
- Location of the forward observer (FO)
- FPLs of all weapons located in each defense area of the platoon
- On-call targets
- Barriers and early warning devices
- Claymore mines

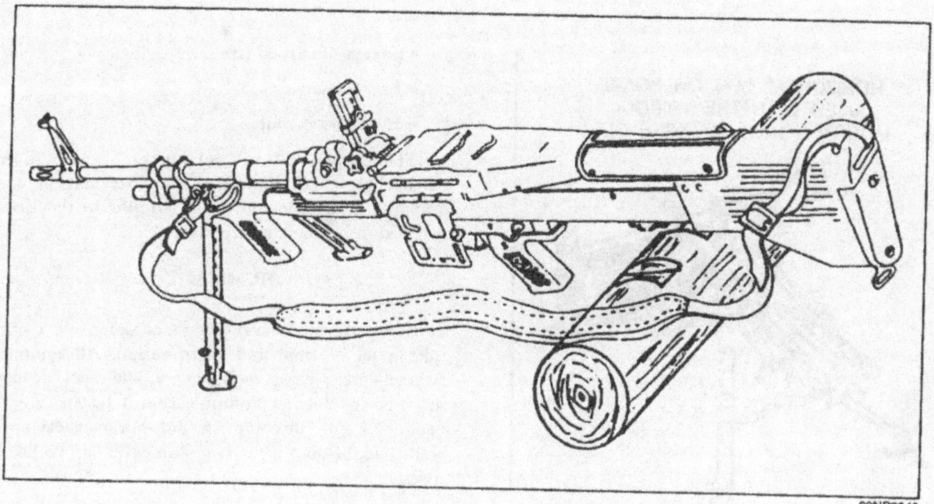

60NP0043

Figure 4-28.—Notched log under receiver.

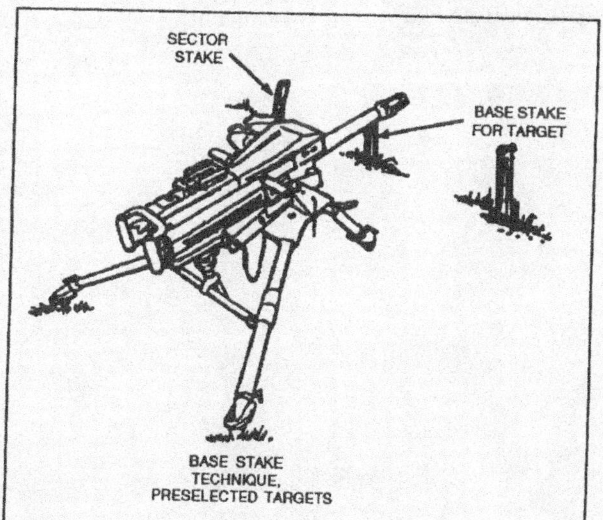

60NP0044

Figure 4-29.—Base stake technique.

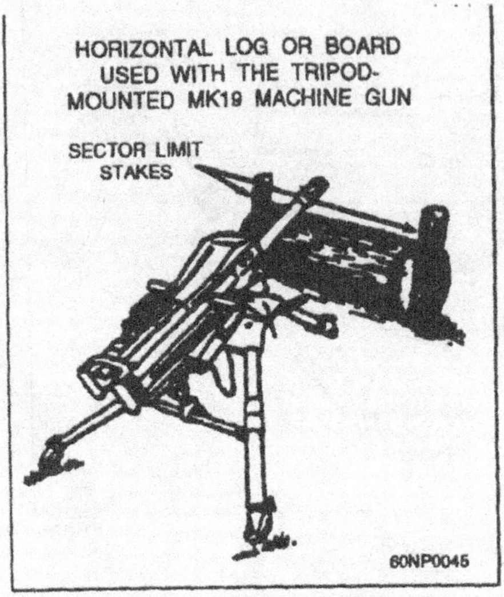

Figure 4-30.—Aiming stakes used with a log.

- Layout of barbed-wire entanglements
- Patrol routes
- Evacuation routes

The company CP will submit the fire plan to the COC. The COC compiles the information and prepares a battalion fire plan similar to the items required in a company fire plan.

SUMMARY

Planning and development of defensive tactics take a lot of time and coordination. All company commanders, platoon commanders, and squad leaders must be involved. Communication between them is essential for developing defensive tactics. A well-thought-out defensive plan will surely be an asset.

"Know the enemy, know yourself; your victory will never be endangered. Know the ground, know the weather; your victory will then be total." Sun Tzu (500 B.C.)

CHAPTER 5

COUNTER AMBUSH TECHNIQUES

Seabees spend a considerable amount of time convoying from the bivouac area to project sites. This makes it important for you to know and understand proper convoy procedures. COMSECONDNCB/COMTHIRDNCBINST. 3122.1 provides the guidance you need for convoy procedures. Another good source for convoy procedures is the FMFM 4-9, *Motor Transport.* This chapter will focus on techniques for use in counter ambush operations that supplement established tactical procedures. It includes counter ambush preparations for motorized units and immediate action drills.

COUNTER AMBUSH TECHNIQUES FOR CONVOYS

Experience in fighting terrorists and guerrillas has shown that they favor the ambush anytime a favorable situation exists. The ambush of units traveling in a motor convoy is not difficult and can be very costly if you do not take time to prepare vehicles and occupants before contact.

CHARACTERISTICS OF A VEHICULAR AMBUSH

A convoy is never safe from a guerrilla ambush. There is no set pattern for a likely area for an ambush. Ravines, defiles, and heavily wooded or jungle-covered areas are most favorable but ambushes are equally likely to be conducted in villages or in flat terrain that offers a minimum of cover and concealment. The most common characteristics of an ambush are as follows:

1. The ambush lasts the minimum time necessary to accomplish the mission.

2. The ambush occurs in two phases—a short period of heavy fire followed by an assault of the ambushed vehicles to capture equipment, to complete the annihilation of personnel, and to destroy vehicles.

3. The basic ambush weapons are small arms. These are augmented by machine guns, rocket launchers, and recoilless rifles.

4. The enemy will use electrically detonated mines to disable vehicles and cause personnel casualties. These mines may consist of artillery shells and mortar rounds as well as conventional mines.

PREPARATION OF VEHICLES

Occupants traveling in vehicles must have all-around observation and fields of fire and can throw or fire grenades without hindrance. They must be able to exit from the vehicle rapidly with minimum restriction. For these reasons the configuration of vehicles, such as a 6 by 6 cargo truck, must be altered. The following measures are taken to "harden" a vehicle and provide its occupants with a degree of protection:

1. Canvas, bows, windshields, and doors are removed.

2. The tailgate is lowered to a horizontal position. A piece of pipe, wood, or metal is affixed to the vehicle in a vertical position extending above the driver's head. This will prevent decapitation from wire stretched across the road.

3. Sandbags are placed on the floorboards and bed of the vehicle. For a 6 by 6 truck, a single row of sandbags is placed on the bed of the truck. Then sandbags are stacked five layers high down each side of the truck. This provides protection from most small arms. A total of 70 to 100 sandbags is required for each truck This load plus troops and equipment weight permits off-highway operation without undue wear on the vehicle. A wooden bench or packs rigged down the middle of the bed provide the troops with seats (fig. 5-1.)

4. Sections of scrap armor plate may be used to reinforce sandbags in the bed of the vehicle.

ORGANIZATION OF OCCUPANTS

Encounters with a guerrilla ambush are sudden, short, and unexpected. The opportunity to inflict casualties upon the guerrillas is lost if your troops are not organized and well-drilled to take immediate

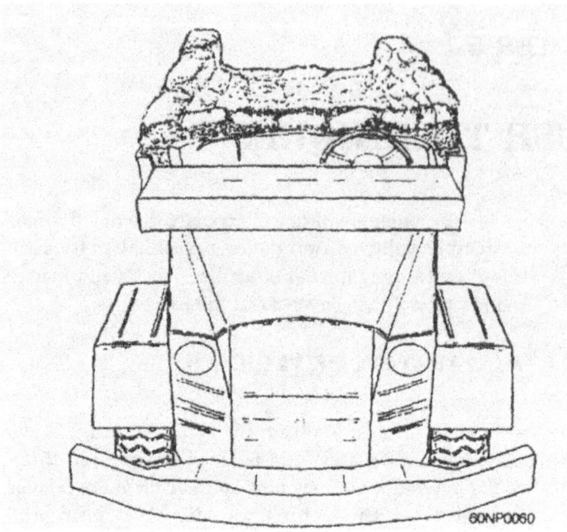

Figure 5-1.—Hardened vehicle.

offensive action. The organization of a rifle squad as occupants of a 6 by 6 truck is as follows:

1. A vehicle commander is designated for each vehicle. The squad leader is usually the vehicle commander and is positioned in the bed of the vehicle where he or she can best control the squad and operator of the vehicle.

2. An assistant driver (shotgun rider) capable of operating the vehicle is seated in the cabin with the driver. The shotgun rider is armed with a M-203 grenade launcher with the M-16 as a automatic rifle. After debarkation, the shotgun rider remains with the vehicle to act as close protection for the driver and vehicle. He or she does not accompany maneuvers executed by the occupant squad.

3. Four comer sentries are positioned in the bed of the vehicle. The two at the front observe an arc of 90 degrees from the front to each side. The two at the back observe an arc of 90 degrees from the back to each side. When possible, each sentry is armed with an automatic rifle. If the vehicle is ambushed, the sentries fire immediately from their positions within the vehicle. Their fire covers the debarkation of the occupants if the vehicle is halted in the ambush killing area. They also help the vehicle commander by notifying the commander of any convoy formation.

4. If a machine gun team is traveling with the occupant squad, it should be positioned facing out the rear of the vehicle and be prepared to debark expeditiously, bringing fire to bear on the enemy and covering the debarkation of the four corner sentries.

5. The remaining occupants are positioned in the bed of the vehicle, each facing outboard.

6. The maximum number of people in the bed of a hardened 6 by 6 truck should not exceed 13 (fig. 5-2).

CONVOY COMMANDER

The position of the convoy commander is where he or she can best control the convoy. This position should never be the lead vehicle. A second in command and a vehicle commander for each vehicle are designated by the convoy commander. Briefing by the convoy commander before departure is detailed and explicit. All drivers and vehicle commanders are present. The briefing includes the following:

- Formation (close column, open column, or infiltration)
- Timings
- Route
- Speed
- Order of march (organization of vehicles and vehicles intervals)
- Communication (radio frequencies, horn signals, and arm signals)

NOTE: Figure 5-3 shows recommended arm signals. It is important that each member of the

Figure 5-2.—Organization of occupants in a hardened 6 x 6 truck.

Figure 5-3.—Convoy arm signals.

convoy know what the arm signals mean no matter which arm signals are used.

- Procedure when communication is lost
- Action on vehicle breakdown
- Actions on ambush
- Action in danger areas
- Call for fire procedures
- Check points
- Location of friendly forces

AMBUSH DEFENSE AND ACTION ON CONTACT

Whatever the precautions and preparations, the ambush is nearly always an unexpected encounter. Counter ambush drills are simple courses of action designed to deal with the problem of the unexpected encounter. They call for immediate, positive, and offensive action. The action on ambush is to drive through the ambush area or stop before running into it, then to attack the enemy immediately from the flank or the rear. When vehicles are fired upon, the following actions are taken:

1. Drivers attempt to drive through the killing zone.

2. Personnel return fire immediately.

3. When vehicles are clear of the killing zone, they are halted. Occupants dismount and take immediate offensive action against the enemy positions.

4. Subsequent vehicles approaching the killing zone halt short of the zone. Occupants debark and take immediate offensive action against the enemy positions.

If hardened vehicles are forced to halt in the killing zone, all available weapons are used to return fire immediately. Occupants remain in the vehicle. On the first perceptible slackening of enemy fire, occupants dismount. When riding in a "soft" vehicle and caught in a killing zone, occupants dismount immediately. In both cases, occupants dismount under the covering fires of the four corner sentries, who initially remain aboard. The occupants then deploy to the side directed by the vehicle commander and take the enemy under fire to cover the dismount of the four sentries.

After dismounting, if no cover is available, an immediate frontal assault against the enemy is employed. The most logical course of action after dismounting is to take cover, immediately establish a base of fire, and use a maneuver element against the enemy ambush positions. Speed of execution is critical.

COUNTERACTION TACTICAL CONSIDERATIONS

The most effective counteraction to ambushes is a flanking attack by elements not in the killing zone quickly followed by relentless pursuit of the enemy. Attention must be given to the following tactical considerations:

1. In actions when no troops have entered the killing zone, the convoy commander launches an immediate flanking attack on the enemy position, using supporting fires from machine guns and mortars.

2. Fire in the killing zone may be from only one side of the road with a small holding force on the opposite side. To contain the convoy element in the killing zone, the enemy will place mines and booby traps on the holding force side. Take care when assaulting the main ambush force because mines are commonly used to protect its flanks.

3. In actions where some portions of the vehicles are ahead and out of the killing zone and the remainder are halted short of the zone, the portion that has not yet entered the killing zone initiates the flanking attack. If the convoy commander is not pressent, the senior vehicle commander takes command and directs the attack. Troops in vehicles which are ahead of the killing zone dismount. Under the command of the senior vehicle commander, the troops return to the vicinity of the killing zone and exploit the situation.

4. The best way an armored vehicle can help in counter ambush action is by moving into the killing zone to engage the enemy at short range. In this way it can give good covering fire to the flanking attack or provide protection for those troops caught in the killing zone.

It is possible that the convoy commander may be killed, wounded, caught in the killing zone, or positioned on the wrong side of the zone. It is essential that all vehicle commanders and squad leaders know their responsibilities for organizing and directing a counter ambush action. This is clearly stated in unit convoy orders and emphasized at briefings. The techniques outlined above are practiced repeatedly until the reaction procedures become a predrilled response, permitting immediate and positive action on ambush.

AMBUSH DEFENSE FOR AN UNBLOCKED ROAD

Guerrillas are seldom able to contain an entire convoy in a single killing zone. This is because of the extensive road space occupied by even a platoon size convoy, and because security or lack of available forces may limit the size of the ambushing force. More frequently, a part of a convoy, either head, tail, or a section of the main body, is ambushed.

That part of a convoy that is in the killing zone and receiving fire must drive out of the ambush if the road to the front is open. Vehicles disabled by enemy fire are left behind or, if blocking the road, are pushed out of the way by following vehicles. Armored escort vehicles must not block convoy vehicles by halting in the traveled portion of the road to return fire. Vehicles that have not entered the killing zone must not attempt to do so. They should stop and personnel should dismount and take defensive positions. Elements of the convoy should not fire on suspected enemy positions without coordinating with the escort forces. The escort vehicles may have left the road in an attempt to overrun hostile positions. Other actions available to convoy personnel for the neutralization of the ambush force are as follows:

1. Direct any vehicles mounted with weapons to lay down a heavy volume of fire on the ambush force.

2. Call for artillery fire on enemy positions.

3. Call for close air support on enemy positions.

4. Call for reaction forces.

5. Direct all nondriving personnel to place a heavy volume of fire on enemy forces as rapidly as possible, as vehicles move out of the killing zone.

A motor transport convoy with a limited escort is seldom able to defeat a hostile force and should not attempt to do so. When part of the convoy is isolated in the killing zone, vehicles that have not entered the ambush area should turn around and return to the nearest secured area until supporting forces can clear the ambush.

AMBUSH DEFENSE FOR A BLOCKED ROAD

When an element of a convoy is halted in the killing zone and is unable to proceed due to disabled vehicles, a damaged bridge, or other obstacles, personnel will dismount, take cover, and return a maximum volume of fire on enemy positions. Security troops from vehicles that have passed through the ambush area dismount and prepare to attack the flanks of the ambush position. Leave a security force behind to protect these vehicles. Personnel in vehicles that have not entered the killing zone follow the same procedure. Before attempting to flank the ambush force, the convoy commander should ensure that the force will not be in the vicinity of any fire missions.

When a security escort is provided and a combat emergency arises, the convoy commander retains operational control unless the command responsibility has previously been assumed by the area commander in whose zone the convoy is operating. **Normally, the security force will take action to neutralize the ambush while the convoy escapes from the killing zone.**

If immediate air or artillery support is available, personnel will be restricted to a specified distance from the road to avoid casualties from friendly fire. In this situation, personnel in the killing zone establish a base of fire, while others take up defensive positions around their vehicles. Everyone waits while supporting fire is called in on the enemy positions.

When the enemy is defeated or has retreated, the road must be cleared and convoy movement resumed as soon as possible. Wounded personnel are evacuated, usually by helicopter. When disabled vehicles cannot be towed, their cargo is distributed among other vehicles if time permits. When it is not feasible to evacuate vehicles and/or cargo, they will be destroyed upon order from the convoy commander. When possible, radios and other critical items will be recovered before the vehicles are destroyed. Under no circumstances will anything in the convoy be allowed to fall into enemy hands.

CONVOY DEFENSE, MINES, AND BOOBY TRAPS

Mines and booby traps are frequently used by ambush forces. Command-detonated mines are often used to start an ambush. Mines will also be planted along the shoulder of the road. This is for harassment and interdiction. A booby trap system may be used against personnel in vehicles. They could consist of hand grenades attached to tree branches over the road where antennas or other projections from vehicles will snag and detonate the grenades. Claymore mines may be suspended from trees and command detonated when a vehicle passes. The following guidelines have proved effective in decreasing damage by mines in convoy operations:

1. Follow the tracks of the vehicle in front.

2. Avoid driving on the shoulder of the road.

3. Whenever possible, do not run over foreign objects, brush, or grass in the road.

4. Avoid fresh earth in the road.

5. Watch local national traffic and the reactions of people on foot. (They will frequently give away the location of any mines or booby traps.)

6. Heavy vehicles, such as tanks, are useful in exploding small mines when deployed in front of the convoy.

7. Wear protective equipment (protective helmets and armored vests).

CONVOY DEFENSE FOR SNIPER FIRE

Exercise caution when sniper fire is received to ensure that any return fire does not harm friendly troops or civilians in the area. The best action to take is passive, and you should ensure that all personnel wear protective helmets and armored vests. **All vehicles should move through the area without stopping.** Escort personnel must notify the march element commander by giving a prearranged signal, such as a smoke grenade, thrown in the direction of fire. Any attempt to locate and destroy the sniper must be long-range fire. The convoy commander may order additional fire or supporting forces into the area to destroy, capture, or drive off the sniper. Convoy personnel should be aware that a heavy volume of fire is frequently used by the enemy to slow down a convoy before an ambush.

CONVOY DEFENSE AGAINST AN AIR ATTACK

The air threat varies from armed helicopters to high-performance aircraft. Convoys face the greatest danger of an air attack while moving along open roads and during halts where there is no overhead cover. An air attack is a type of ambush; therefore, many procedures used during a ground ambush are also applicable to the air attack. For example, the convoy commander must do the following:

1. Prescribe alarm signals.

2. Give instructions for immediate action to take when under attack.

3. Prescribe actions to take in the absence of orders.

4. Make sure that defense procedures are rehearsed.

5. Review the procedures with convoy personnel before the convoy moves out.

The convoy commander must remember that enemy pilots may do the following:

1. Tries to surprise the convoy.

2. If aircraft attack from higher than 350 meters, small-arms fire will not have much of a chance against them but air defense weapons can be used against them.

3. Will fly at high speed, using high-G maneuvers and evasive action to make air defense weapons and small-arms fire less effective.

Lookout Procedures

Air guard duties are assigned to each individual throughout the convoy, and each person is given specific search areas (scanning patterns). As much as vehicle configuration will allow, search areas assigned should cover a full 360 degrees in overlapping sectors. If the road march lasts more than an hour, Seabees should take turns at air guard duty. Scanning for a long period (over 30 minutes) dulls the ability to spot aircraft.

Camouflage and Concealment

Camouflage and concealment techniques are used to make it more difficult for the enemy to spot the convoy. Not much can be done to change the shape of a vehicle moving down the road, but the type of cargo being transported can be disguised or conceded by covering it with a tarpaulin. Bulk fuel transporters (tankers/refuelers) are usually priority targets. By rigging tarps and bows over the cargo compartment, you can conceal the nature of the cargo from the enemy pilot. Other effective cover and concealment measures are as follows:

1. Operators must be trained to break the shape of the vehicle as seen from the air by looking for a bush, a tree, or other means of concealment as they disperse.

2. Smooth surfaces and objects (windshields, headlights, and mirror) will reflect light and attract the attention of the pilot. All shiny items should be camouflaged or covered before the convoy moves out (fig. 5-4).

If vehicles are not painted in a pattern to blend with the terrain and to break the outline, you can use mud to achieve this effect.

Communications Security

Communications equipment can be very useful for controlling convoys. But it can also help enemy pilots find you. Use the radio when necessary, but be brief.

ACTIVE DEFENSE AGAINST AIRCRAFT

The convoy commander may choose between an active and a passive defense against the air threat. In an active defense, the amount of fire a logistic convoy can bring to bear on attacking aircraft is usually limited. It is limited to the number of vehicles with mounted machine guns and the individual weapons of vehicle operators and assistant operators. However, convoy personnel should not be led to think that trying to shoot down an attacking airplane with small-arms weapons is fruitless. In the Korean conflict, the U.S. Air Force lost 544 aircraft to combined small-arms and air defense fire. Over North Vietnam, small arms contributed to significant losses of U.S. aircraft. During the Middle East War in 1973, units on both sides used small-arms weapons to drive off, damage, or destroy attacking aircraft.

NOTE: The key to effective small-arms fire against aircraft is volume. Put up a large volume of fire with small caliber weapons. Accuracy is not important; volume is!

Figure 5-4.—Cover portions of vehicles that reflect light.

Firing Positions for Small Arms against Aircraft

Except for the prone firing position, the basic firing stances of the rifleman remain the same. A rifleman quickly learns that he or she wants to fire from some type of cover or concealment; therefore, you should look for a tree or a large rock to help support the weapon and provide some protection.

The M-60 machine gunner should also fire from a protected position if possible. In a real emergency, another Seabee can act as a hasty firing support for the machine guns. The M-2 .50 caliber machine gun can only be used from the ring mount of dedicated vehicles. The following are tips for Small-Arms Air Defense:

1. Fire at any attacking aircraft using all available weapons.

2. Fire at the nose of an aircraft.

3. Volume of fire is the key; everybody fires.

4. Lead aircraft crossing your position (M-16, M-60, and M-2 should lead jets by the length of one football field).

5. Take cover if you have time.

6. Support your weapon, if possible.

7. Lie on your back if caught in the open so you can fire upon the aircraft.

8. Mounted M-60s and M-2s should aim slightly above the nose of the aircraft for head-on targets.

Passive Measures

For a logistic convoy without significant firepower, passive measures are most effective. The key here is to prevent attacks by hostile aircraft. Dispersion is the key for air defense. The formation used by the convoy is a type of passive defense. The convoy commander must decide whether to use an open, closed, or infiltration column. Distance between vehicles must be flexible. It should vary from time to time during a march. Factors influencing selection of the best vehicle distance include the following:

- Mission
- Cover and concealment along the route
- Length of the road march
- Type of road surface

- Types of vehicles
- Nature of cargo
- Enemy threat (ground or air)
- Presence of defense support
- Small-arms potential

The following are types of formation a convoy commander may choose for passive air defense:

1. **Close Column.** Each vehicle follows the vehicle ahead at a distance sufficient to ensure against collision. The gap specified is dependent upon road conditions, weather conditions, terrain, and convoy security requirements. Close column formation eases column control and intracolumn communications. During daylight hours, fewer guides, traffic escorts, and route markers are needed. The close column is generally used at night because air attack threats are less likely. It is also used over poorly marked routes when visual contact between vehicles is essential and in areas where hostile action is not imminent. Seabees found this formation to be a success in crowds during Operation Restore Hope in Somalia, Africa. The close column prevented crowds from coming between the convoy and disrupting the movement of the convoy.

2. **Open Column.** In open column formations distances between vehicles are increased to enhance dispersion. This formation offers an advantage in that fewer vehicles are likely to be damaged from air-delivered rockets, cannons, or cluster bomb units. An open column increases the degree of passive protection against hostile observation and attack. It

TYPE FORMATION	WHEN USED	DENSITY PER MILE/KM	SPEED mi/h/km/h	ADVANTAGES	DISADVANTAGES
Close	Night, poorly marked routes, or in congested areas.	67/40	10/16	Full traffic capacity of road can be used. Control is better. Fewer guides, escorts, and route markers are needed.	Quick dispersion is difficult. The column is easily detected. May cause congestion at point of arrival. Requires careful scheduling and control to avoid blocking at intersections. Causes driver fatigue.
Open	Daylight	20/12	15/24	Less chance of enemy observation or attack. Cargo moves faster. Driver fatigue is reduced. Fewer accidents, very flexible.	Command and control are difficult. Proper vehicle spacing is hard to keep.
Infiltration	Daylight, congested areas. Heavy traffic crosses route.	10 or less per hour	Various	Provides maximum security and deception. High speeds are possible. Other traffic has little effect on individual trucks. Does not hinder cross traffic.	More time required to complete the move. Column control is nearly impossible. Drivers can get lost. Specific details must be provided each driver. Maintenance, refueling, and messing are difficult to arrange. Vehicles may bunch up, causing close columns to form. Requires experienced drivers. Orders are not easily changed. The unit cannot be redeployed as a unit until the last vehicle arrives at destination.

Figure 5-5.—Types of column formation.

permits greater highway speeds with safety and provides for greater flexibility in highway use. However, open columns make control more difficult for the convoy commander when it is necessary to give orders to stop, to continue, to disperse and seek concealment, or to engage aircraft. Moreover, the column is more likely to be attacked since it is exposed for a longer period of time. If attacked, its defense is less effective because small-arms fire is less concentrated.

3. **Infiltration.** In infiltration formations, vehicles are dispatched as follows:

a. Individually

b. In small groups

c. At irregular intervals

d At a rate that minimizes changes in the average traffic density and prevents massing of vehicles

Average distance between vehicles and groups is decided by the rate at which vehicles are dispatched. Deception is provided by intermingling various types of vehicles and by permitting passing within the column. Infiltration may provide the best passive defense against hostile observation and attack; however, it provides the least active defense capability if individual or small groups of vehicles are attacked. This method permits individual vehicles to travel at high speeds and cause less cross traffic interference. It permits use of a route on which heavy traffic normally precludes the entire unit moving at one time (fig. 5-5).

PASSIVE REACTIONS TO AIR ATTACKS

Seeing the enemy first has long been established as an element of survival in any combat situation. The advantages of seeing enemy aircraft first is largely self-evident, and the importance cannot be overemphasized. All convoy personnel must be thoroughly briefed as to their responsibilities in the employment of effective scanning techniques. When aircraft are spotted or early warning is received, the convoy commander has three options:

- Continue to march.
- Have the convoy immediately stop in place.
- Disperse quickly to concealed positions.

Continue to March

The mission and/or terrain may dictate that the march continue. If this is the case, the convoy speed should be increased. Continuing the march offers the advantage of preventing a moving target and making it more difficult for the enemy to hit. Disadvantages exist in that detection is easier and volume and density of small-arms fire are reduced.

Stop the Convoy

If the convoy commander chooses to halt the convoy, the vehicles simply pull to the shoulders of the road. This technique has several advantages:

1. Harder for the pilot to see the convoy.

2. Easy to continue the march.

3. Volume and density of organic weapons will be greater than if the convoy disperses.

However, a disadvantage exists in that a convoy stopped on the open road makes a good target for enemy troops. The enemy has a better chance of causing serious damage.

Disperse the Convoy

A simple technique that a convoy commander can use in dispersing vehicles is to establish a method in Standard Operating Procedures (SOPs). SOPs will emphasize that in case of an air attack odd-numbered vehicles go to the left, and even-numbered vehicles go to the right. The key to dispersion is not to make two straight lines out of what was one long line; the vehicles must be staggered. This should not be a problem if the drivers have been trained to go to trees, bushes, and folds in the ground that will give concealment (fig. 5-6).

Once the convoy is dispersed, all personnel, except vehicular mounted weapons gunners, dismount and take up firing positions. Advantages of this system are that it is more difficult for the enemy pilot

Figure 5-6.—Vehicles dispersed and in concealed positions.

to detect the vehicles and get multiple hits. However, this method has several disadvantages:

1. Easier for the enemy pilot to spot the convoy as it begins to disperse.

2. Volume and density of small-arms fire are reduced.

3. Takes longer to reorganize the convoy after the attack.

ARTILLERY OR INDIRECT FIRE

Enemy artillery units or indirect fire weapons may be used to destroy logistic convoys or to harass and interdict the forward movement of supplies and personnel. Artillery fires are either preplanned fire or fires called in and adjusted on a target of opportunity by a forward observer. Of the two, the adjusted fires present the more complex problem as the artillery barrages can be adjusted to follow the actions of the convoy.

Active Defense

Active defensive measures against artillery are limited but must not be overlooked. It is important that several personnel are highly capable of calling in a fire mission. Active measures would include the following:

1. Directing counterbattery fire if the direction and approximate distance to the enemy artillery can be estimated.

2. Directing small-arms fire or artillery fires against the enemy forward observer if they can be located.

3. Coordinating air strikes against the enemy artillery.

Passive Defense

The convoy commander has three passive defense options when confronted with incoming artillery rounds. The options are as follows:

- Halt in place.
- Continue to march.
- Disperse quickly to concealed positions.

Regardless of the option selected, the actions to be taken and the signal directing the action should be covered in the unit SOP.

The convoy should only be halted when the artillery concentration is ahead of the convoy. The convoy commander should look for an alternate route around the impact area and the convoy should remain prepared to move out rapidly.

The missions and/or terrain may require the convoy to continue. If this is the case, increase speed and spread out to the maximum extent the terrain will allow. Casualties can be reduced by the following:

- Avoiding the impact area
- Increasing speed
- Increasing dispersion
- Wearing individual protective equipment
- Using the vehicle for protection

IMMEDIATE ACTION DRILLS

The guerrilla normally seeks contact with organized units only under favorable tactical circumstances; for example, ambushes. When contact is made under less favorable circumstances, the guerrilla attempts a rapid withdrawal. In either case, small-unit encounters with guerrillas are likely to be sudden, violent, and of short duration. Slow reactions to an ambush can result in excessive losses or the loss of an opportunity to punish the guerrilla unit. Contact is often made at close range, particularly when operating in jungle, temperate zone forests, woods, or heavy brush. Immediate action drills aid small units in reacting quickly and properly.

Immediate action drills are predrilled, prerehearsed reactions to contact or anticipated contact with the enemy. Immediate action drills are most frequently used by rifle platoons and squads. They are used during the conduct of foot patrols and dismounted movements in close terrain against guerrillas.

The variety of drills is limited only by the imagination and initiative of the unit leader and the state of training of the unit. It is impractical to develop drills covering every contingency; however, it is important to develop a drill for each of the most frequently occurring situations. The response to a given situation must not be stereotyped, as the enemy may ultimately capitalize upon rigid adherence to the same tactics.

Immediate action drills stress simplicity, aggressiveness, and rapid execution. They demand alertness and a high state of individual training. Drills are of little value to a unit in which the individual Seabee lacks proficiency in the fundamental combat skills.

VEHICLE UNLOADING DRILL

When a vehicle is forced to halt in the killing zone of an ambush, the debarkation of occupants must be organized and predrilled. On order or signal, the response must be immediate and each person must act swiftly to move to the proper position. Confusion is thus overcome and immediate offensive action against the enemy is more likely to be effective. When the vehicle is halted, the actions are as follows:

1. If the vehicle is hardened, the vehicle commander acts approxiately as previously discussed. The vehicle commander then commands "DEPLOY RIGHT (OR LEFT)," to show the direction in which the occupants are to assemble after dismounting.

2. Sentries throw smoke grenades and open fire immediately on the ambush positions. The grenadier, if one is aboard the vehicle, fires on the ambush position.

3. Occupants, under cover of fire from the sentries, dismount over both sides of the vehicle and move to the side of the vehicle indicated in the command. As few occupants as possible attempt to dismount over the tailgate of the vehicle.

4. When the occupants have dismounted, the sentries dismount under covering fire from troops on the ground.

5. The driver and assistant driver dismount in the direction indicated by the vehicle commander.

6. When all occupants are out of the vehicle, action is taken as previously mentioned in this chapter.

FREEZE AND HASTY AMBUSH DRILL

The freeze and hasty ambush is a drill designed to deal with the meeting engagement. The drill is undertaken when the unit has sighted guerrillas approaching but has not yet been seen by them. Immediate action is taken to ambush the guerrillas when their approach is moving on a trail different or the same used by the unit.

1. Freeze—When the guerrillas are sighted, the unit is halted by silent signal, such as an arm and hand or other prearranged special signal. The signal is relayed to each member of the unit, and each member freezes in their tracks with their weapons in the firing position.

2. Different trails—If the guerrillas are approaching on a route different from that of the Seabee unit, the unit remains on the trail in a freeze position. The unit leader signals Commence Firing when the guerrillas present a suitable target.

3. Same trail—On initially sighting the enemy, the freeze is executed. The individual making the sighting indicates the number of enemy by silent signal and then moves off the trail. Each individual relays the signal and moves off the trail on the same side used by the originator. It is essential that the entire unit move to the same side of the trail. Speed of execution and silent movements are mandatory. Any unnatural sound may cause the guerrillas to turn and flee. Each member of the unit takes up a firing position facing the direction of enemy approach. The unit leader initiates the ambush by firing his or her weapon. In the event the guerrillas sight and fire upon a unit member other than the unit leader, that individual fires and springs the ambush.

IMMEDIATE ASSAULT DRILL

The immediate assault is a tactic used during an unexpected encounter at close quarters. It is a predrilled response to situations in which the guerrillas and the Seabee unit become aware of each other simultaneously. The immediate assault drill is a rapidly executed frontal assault.

1. The drill is usually initiated by the first member of the unit who sights the enemy. He or she fires at the enemy and shouts a prearranged signal showing the direction of the encounter; for example, "enemy front (left, right), charge."

2. The signal is repeated by each individual.

3. The unit adopts the line formation, oriented in the indicated direction of contact. A predesignated subordinate unit is withheld from the line to protect the flanks and rear.

4. The unit leader sounds a prearranged assault signal.

5. The assault is pressed forward until halted by the unit leader, usually when the guerrillas are no longer insight.

COUNTER AMBUSH DRILL

There is no generally accepted immediate action for foot troops when ambushed. Adherence to the principle of security in avoiding an ambush is easier than to escape from one. When ambushed, violent and concerted reaction is required to prevent annihilation. Units must have a prearranged plan, known to every troop, which allots a specific immediate action to each individual according to his or her location and function in the formation.

Entire Unit in Killing Zone

It is seldom possible to find covered or concealed positions within the killing zone from which to exchange fire with the enemy. The unit may execute the immediate assault in the direction shown by the unit leader. The direction shown is normally the weakest point in the ambush and is a prearranged counter ambush drill.

The unit may initially execute a preplanned movement to a position outside the killing zone indicated by the unit leader. Normally, the position is one that provides cover and concealment and is the location from which a subsequent drill is undertaken to eliminate the enemy.

Leading Element in Killing Zone

When only the foremost elements of the unit are caught in the killing zone, an immediate encircling attack is executed and actions are taken as follows:

1. Elements within the ambush indicate the nature and location of the ambush by prearranged signal; for example, by voice "ambush front (left, right)" or by whistle or other signals.

2. Previously determined ambushed elements, execute an immediate assault.

3. Previously determined base of fire elements from personnel not yet engaged in the ambush, assume base of fire positions. These fires simultaneously support encircling maneuver elements and the personnel in the killing zone.

4. The encircling attack units move out in a prearranged envelopment of the enemy flank and/or rear (fig. 5-7).

NOTE: Take care in assaulting the main ambush force, as mines are commonly used to protect its flanks.

5. Whenever possible, the enemy rear is enveloped. The assault by the encircling units drives the enemy into the fires of the base of fire elements.

SUMMARY

The company commander must train his or her troops in convoy defensive techniques. OPSEC measures must be considered during the planning process and practiced during day-to-day hauling operations. The primary types of engagement likely to be encountered during convoy movements are attacks by air, artillery, sniper, ambush, and NBC. The convoy commander must ensure that subordinate commanders and drivers are aware of and trained on

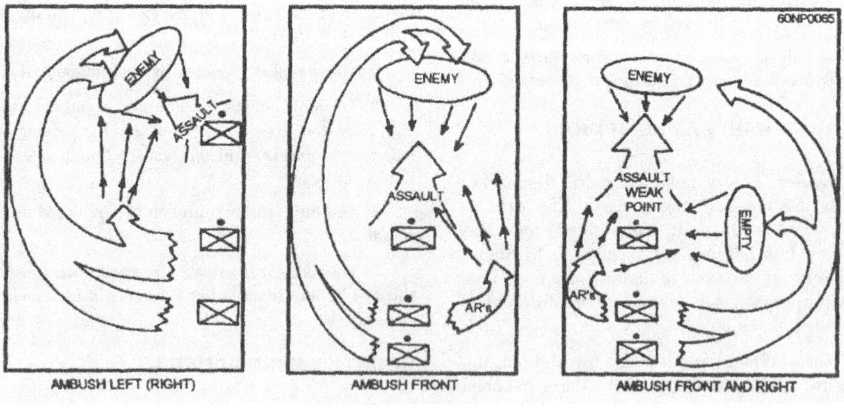

Figure 5-7.—Encircling attack.

the procedures to be used for countering the different type of attacks. Active and passive defense techniques must be learned and practiced. In an ambush situation, immediate reaction and aggressive leadership are essential to limit casualties and damage to vehicles, cargo, and personnel.

The following is a list of ambush instructions that was developed from lessons learned during the Vietnam War.

Before Ambush

1. Be alert for changes in familiar scenes along the route.

2. Expect ambush upon detonation of mines, automatic weapons fire, and heavy sniper fire.

During Ambush

1. Notify security force by radio call using prearranged signals. (Do not tie up the radio nets with unnecessary conversation. Maintain strict radio discipline.)

2. Continue moving, maintaining vehicle distance if possible.

3. Place any tracked vehicles in front to activate further detonation of road mines.

4. If disabled, steer off the road to allow passage of other vehicles.

5. If disabled and the convoy is moving through the killing zone, mount a passing vehicle.

6. Do not enter the killing zone when it can be avoided.

7. Provide flanking support fire into the killing zone.

8. Security vehicles apply maximum base of fire; fight as a team; deploy upon command.

9. Prime targets are enemy automatic weapons positions, enemy mortar, rocket positions, and assaulting ground forces.

After Ambush

1. Protect and care for wounded; evacuate wounded as soon as possible.

2. After contact has been broken, reopen roadway, ensuring that it is free of mines.

3. Do not attempt to remove mines; contact explosive ordnance disposal.

4. Provide flank security while the convoy passes through the area.

5. When contact is broken, do not give chase.

6. Stay alert for a counterattack

7. Remember details.

CHAPTER 6

CHEMICAL, BIOLOGICAL, AND RADIOLOGICAL DECONTAMINATION

The presence of contamination can greatly reduce the effectiveness of the combat forces. Contamination forces personnel to wear protective equipment that degrades their ability to conduct individual and collective tasks. Therefore, an understanding of the behavior and characteristics of contamination enables personnel to better direct their efforts in taking countermeasures to avoid or reduce a nuclear, a biological, and a chemical hazard. Consideration of these factors will help the individual Seabee, planner, and leader in the integration of CBR defensive measures in tactical operations.

Companies in a battalion are required to have at least one six-man team trained to perform decon operations. Obtaining the required skills through training at the 20th NCR or 31st NCR is strongly encouraged to become decon qualified. Therefore, this chapter is only designed to familiarize the reader with decon operations and decon terms.

NOTE: Nuclear, biological, and chemical warfare (NBC) and chemical, biological, and radiological warfare (CBR) are similar. Army field manuals (FMs) and Marine Corps field manuals (FMFMs) refer to CBR as NBC; therefore, when this chapter refers to CBR, it is also referring to NBC.

FORMS OF CONTAMINATION

Successful decontamination (decon) requires that you understand the forms of contamination and what makes it dangerous. The different origins and forms of contamination cause different hazards. A brief discussion of the major forms and origins of contamination will clarify the meaning of contamination, as it is used in this chapter.

SOLIDS

Chemical agents, biological agents, and radiological contamination can all take solid forms. Often, they appear as fine dust. The dust could take the form of radioactive dirt (fallout), a frozen chemical mist, or a dust pollinated with biological toxins and/or biological spores. Another form is a powder, coated with chemical/biological agents (for example, dusty mustard).

LIQUIDS

Liquid contamination is generally delivered in a mist, a vapor, or rain that falls to the ground. Liquid contamination can be made thick, like syrup or gelatin. It sticks to what it touches and evaporates slowly. Low vapor pressure and high viscosity make it difficult to decontaminate. Chemical, biological, and nuclear contaminates, when mixed with rain, can contaminate large areas. When the "washout" evaporates, the solid or liquid contamination may remain for some time.

GASES

Chemical contaminants give off vapors. Toxic chemical agents, delivered as a gas cloud, are used either directly on the target or upwind of the target. Depending on the weather conditions, wind currents can spread toxic gas clouds over a large area. Most toxic gases disperse or evaporate quite readily. However, surfaces, contaminated with liquid chemical agents, may give off toxic gases for days.

TYPES OF CONTAMINATION HAZARDS

If you understand the contamination hazards (transfer, spread, vapor, desorption, and radiation), you will understand contamination characteristics. This will help you to understand the importance of decon in successfully completing your mission under CBR conditions.

TRANSFER

Anything that touches a surface covered with liquid or solid contamination will tend to pickup that contamination and move it from one surface to another. You must eliminate or limit contamination transfer into clean areas. For example, troops climbing in and out of a contaminated vehicle, can transfer agents to the inside of the vehicle. This activity results in two hazards: inhaling the gas given off by the agent brought into the vehicle and physical contact with the agent brought into the vehicle.

SPREAD

Touching a surface covered with a liquid chemical agent can spread contamination on the same surface, thus increasing the size of the contaminated area. When this condition occurs, more decontaminates along with more of an effort will be spent deconing. Therefore, limit the spread of contamination to a clean surface by decontaminating it with a designated decontaminate and/or the appropriate equipment.

VAPOR

A vapor hazard includes any contamination you can breathe, no matter what form it takes, such as dust in the air, atomized liquids (aerosols), or true gases. Generally, vapors in an open/outdoor area disperse rapidly so you do not need to decontaminate them. However, some agent vapors, such as atomized blister, create a transfer hazard because they settle from the air and coat the surfaces they touch. Since solid or liquid contamination remains on a surface, it can continually generate new vapors. Liquid contamination, mixed with dust, can result in a vapor hazard due to wind or movement of vehicles. Generally, when a transfer or spread hazard exists, a vapor hazard could also exist.

DESORPTION

Liquid chemical contamination quickly absorbs into porous surfaces. Once absorbed, it begins to desorb or give off gas; that is, low levels of vapor pass out of the contaminated surface into the air and can be transferred to any surface that contacts it, including bare skin. For example, if you were operating a vehicle that was desorbing a nerve agent, you should protect yourself by wearing, as a minimum, your protective mask and gloves. Exposure to the desorbing nerve agent might blur your vision or interfere with your ability to think clearly. Handling a steering wheel bare-handed when it is desorbing nerve agent may also cause acute nerve agent poisoning. Prevent desorption by decontaminating quickly before any agent can be absorbed into the surface. Surfaces, protected with a chemical agent resistant coating (CARC) that consists of a polyurethane paint coating, can prevent agent absorption. These surfaces can be easily decontaminated with soap and hot water or DS2.

RADIATION

The penetrating energy of radiation does not directly fall into any of the previous categories. Radiation is given off by radioactive dust or dirt, most of which appears as fallout. For decon purposes, radiation can be thought of as a solid. Radioactive contamination can usually be removed by brushing, wiping, or shaking. Decontaminate quickly to decrease the cumulative effects of radiation; otherwise, small but frequent exposure to radiation may cause radiation sickness.

PERSISTENCY

The length of time a hazard remains depends on the "persistency" of the contamination. A full discussion of detection and hazard prediction for all types of contamination is given in Army manuals FM 3-3 and FM 3-3-1.

CHEMICAL

Nonpersistent contamination generally requires no decon. However, the duration and effectiveness of chemical agents used on the battlefield will depend on a series of factors that affect agent persistency. Some of these factors are as follows:

- Type of contamination
- Contamination density and droplet size
- Temperature
- Wind speed
- Sunlight
- Humidity and rain
- Composition of the contaminated surface
- Type of soil and terrain

Any contamination found on your skin must be decontaminated immediately, regardless of persistency. Some contamination hazards can affect you within minutes after touching your skin (an agent like CX will affect you within seconds). After you conduct skin decon, use detection equipment to determine the type of contamination. This will help to decide whether additional decon and/or treatment is required.

Changes to the physical behavior of chemical agents can be caused by changes in weather conditions. For example, in cold weather, nonpersistent agents tend to become semipersistent, lasting from 2 to 10 days. Refer to FM 3-6, *Field Behavior of NBC Agents,* for detailed information.

BIOLOGICAL

The many variables, involved in estimating persistency of biological hazards, require separate consideration for each instance of contamination. Specially trained medical personnel will consider specific treatment direction only after the contamination has been specifically identified. For example, biological agents will persist longer in cold weather. Temperate inversions (stable conditions) that exist over snowfields also tend to prolong the stay of an aeorsolized biological cloud.

RADIOLOGICAL

A general idea of the persistency of radiological hazards can be gained by taking radiation dose readings. At this time, the NCF uses two types of radioactive detection and computation (RADIAC) instruments:

1. The AN/PDR-27 instrument provides radioactivity detection in the lower ranges. It is designed to detect beta radiation and measure and detect gamma nuclear radiation. The AN/PDR-27 is also used to monitor low level radiation contamination on personnel, supplies, and equipment.

2. AN/PDR-43 instrument provides detection of radioactivity in the high range. It is designed to detect beta radiation and measure and detect gamma nuclear radiation.

NOTE: Only qualified personnel are authorized to operate RADIAC instruments.

For operational purposes, you can use the "7-10 rule of thumb" to estimate future radiation levels. **This rule provides a general estimate and should be used for planning only.** The rate of radioactive decay is proportional over time. The 7-10 rule means that for seven multiples of time after the burst, the radiation intensity will decrease by a factor of 10. For example, if 2 hours after the burst your radiation reading is 100 Centigray (cGy) (rad) per hour, then 14 hours after the burst (7 times 2 hours), you can expect a reading of about 10 cGy (100 cGy divided by 10). Radiation contamination is not affected by climatic conditions or other variables that affect chemical contamination. The Army manual, FM 3-3-1, describes radiation decay rates in detail.

NEGLIGIBLE RISK

You must consider decon if the levels of contamination exceed negligible risk levels.

BIOLOGICAL AND CHEMICAL

Negligible risk levels for biological and chemical contamination are contamination that will cause mild incapacitation among no more than 5 percent of unprotected troops operating for 12 continuous hours within 1 meter of contaminated surfaces. Measurements that determine safe levels are made with detection equipment held 1 inch away from the surface. For example, a one bar reading displayed on the chemical agent monitor (CAM) indicates a reduced hazard level that should be considered as a negligible risk level.

RADIOLOGICAL

Negligible risk levels for radiological contamination are measurements of 0.33 centigray (cGy) or less. This level of radiation will cause no more than 2.5 percent mild incapacitation to unprotected troops.

CONTAMINATION COMBINATIONS

Simultaneous enemy CBR attacks will probably be part of the strategy of the enemy. Risk assessments include consideration that the enemy may use combinations of nuclear, biological, and chemical weapons or may use any of these combined with conventional fire. Once CBR weapons have been introduced on the battlefield, the enemy may try to deceive you regarding the type of hazard.

The thermal effects of a nuclear blast might destroy the effects of any chemical or biological weapons used simultaneously. However, chemical or biological weapons effectiveness is increased if used following a nuclear attack. Nuclear blast casualties and psychologically stressed troops are vulnerable to a CBR agent attack. Agents could enter collective protective shelters, communication facilities, and vehicles damaged by the nuclear detonation.

When CBR contamination hazards exist, **decontaminate the chemical agents first.** Chemical agents are normally the most lethal and fastest acting type of contamination. The decon methods for chemical agents are also effective for neutralizing or removing biological and radiological contamination. The reverse is not true.

The enemy may use a mixture of agents in their munitions to cause multiple types of contamination. Such mixtures could be used to achieve the following various purposes:

1. Lower the freezing point of the agents and increase agent persistency, such as mustard lewisite mixture.

2. Create both percutaneous (through the skin) and inhalation hazards, such as thickened GD and GB.

3. Complicate agent identification of mixed agents, making treatment of casualties more difficult.

4. Combine agents with both immediate and long-term persistency, such as anthrax, with an incubation period of 1 to 5 days and histoplasmosis (pulmonary infection disease) with an incubation period of 5 to 18 days.

There is no field detecting system that can detect or identify biological agents. Therefore, combinations of biological and chemical contaminants present a different challenge. This challenge can be dealt with if standard chemical decon measures are followed at once. Use standard chemical decontaminants when combinations are known or suspected to exist. They can be used for toxins and biological agents as well as chemical agents. See appendix IV for a description of field expedient chemical decontaminants for use against these hazards.

Do not base decon measures solely upon the first hazard identified. Make sure you check thoroughly to identify all agent hazards. When specific agents are detected, take appropriate decon measures.

REASONS FOR DECON

You must have a good idea of the reasons for decon and the types/techniques of decon. You must assess your tactical situation and consider your decon resources within the context of mission, enemy, terrain, troops, and time available (METT-T). You must know the principles of decon and know how decon affects your combat power. Protective clothing and equipment (MOPP gear) and collective protection shelters offer only a temporary solution. Decon is a more permanent solution because it includes the removal, the destruction, or the neutralization of contamination. When you have become contaminated, there are practical reasons why at least some decon must occur as soon as possible. You must use these decon concepts in selecting the best action to take to accomplish your mission.

Decon should be considered within the context of METT-T and resources available. The four factors that must be addressed before you decide to decontaminate are as follows:

- Lethality
- Performance degradation
- Equipment limitations
- Transfer and spread

LETHALITY

Some types of contamination are so toxic they can kill or incapacitate if they contact exposed skin for a few minutes. If your skin becomes contaminated, you must stop breathing, mask, give the alarm, and decontaminate your skin immediately. Periodically, observe for nerve agent symptoms if the agent type is unknown.

PERFORMANCE DEGRADATION

MOPP (mission-oriented protective posture) gear provides protection but also degrades performance. The longer you are in MOPP 4, the lower your efficiency. Using tools and weapons while wearing the protective gloves is awkward. The mask reduces your field of view, making it difficult to use some optical sights and night vision devices. Extended operations in MOPP gear tire and discourage troops. Troops cannot eat while wearing a mask. Urinating and defecating are potentially dangerous in a contaminated area. Even resting and sleeping are difficult because it is hard to breathe through the mask. Hot or humid climates compound these problems because the mask makes breathing more difficult.

EQUIPMENT LIMITATIONS

MOPP gear provides protection from chemical and biological agent attacks, but some limitations can reduce its effectiveness. Agents can gradually penetrate the mask hood. However, the protective qualities of the hood can be extended many hours if the chemical agent is removed or decontaminated quickly before it penetrates the hood.

Two primary types of chemical protective overgarments are in use at this time with different protection. The first is the chemical protective overgarment (CPO) that is currently in the TOA of each battalion. In an uncontaminated environment, the CPO is good for 140 hours within 30 days after the initial opening of the original protective package and subject to the following conditions:

- When not worn, store in a sealed plastic bag.
- The 140 hours is only applicable to wear time. The clock stops while the overgarment is stored in a plastic bag.
- Overgarments that have come into contact with POLs must be replaced.

In a **contaminated** environment, 6 hours of protection is provided and the CPO is subject to the following:

1. Once in a contaminated environment, the 6-hour criteria vice 140-hour criteria becomes the controlling parameter. If chemical liquid contamination remains after 6 hours, MOPP gear exchange must be conducted as soon as possible. "MOPP Gear Exchange" will be discussed further on in this chapter.

2. If cumulative wear time is 138 hours and the environment becomes contaminated, the suit may be worn an additional 6 hours for a maximum cumulative wear time of 144 hours.

The second type of overgarment that is currently not in the TOA is the battle dress overgarment (BDO). It provides protection for 24 hours after becoming contaminated with liquid chemical agents.

Troop performance will decrease over time while they are in full MOPP; therefore, risk assessment must be conducted before executing a task or mission. For example, you should know that the following critical areas or tasks are degraded:

- Navigation and terrain orientation
- Target acquisition
- The decision-making process (leader fatigue)
- Communications
- Fire support coordination
- Maneuver formations/convoy operations

Unit commanders, conducting extended operations in full MOPP gear, weigh the risk of whether or not to conduct decon. The unit commander must consider the capabilities and limitations of the unit while performing in MOPP during different types of conditions. Refer to the *Seabee Combat Handbook*, volume 1, chapter 19.

A contamination hazard will be reduced by conducting hasty decon and exchange of the MOPP gear. This action also provides temporary relief from MOPP and the reduction of risk during combat operations. The commander considers the time and the resources needed to conduct decon versus the degradation caused by operating in full MOPP gear. The commander also understands that before ordering the removal of protective masks, they must move their unit to a clean area and conduct unmasking procedures. After hasty decon (MOPP gear exchange and vehicle wash down), the Seabee's risk is reduced by the following:

1. Decreasing the time personnel are exposed to chemical agents.

2. Providing temporary relief from full MOPP. This aids in increasing survivability of the unit on the battlefield because the MOPP encapsulation causes limitations that may result in conventional casualties (for example, heat stress in combat).

3. Decreasing the risk of transfer and spreading of contamination.

MOPP gear provides little direct protection from the hazards of radiological (rad) contamination, such as the radiation from fallout. But commanders may often decide to use MOPP gear for its indirect advantages. MOPP 4 can prevent the inhalation of radioactive particles, keep contamination off the skin, and greatly simplify decon. Although the danger from fallout is not immediate, the radiation may gradually build up to a dangerous level. Therefore, the contamination must be removed when the mission allows. The protective mask filter elements will need to be replaced because of radioactive particle buildup.

TRANSFER AND SPREAD

You must avoid contamination as much as possible. Once a unit becomes contaminated with a chemical agent, quick or rapid decon is critical to prevent further spread or transfer of contamination onto a clean area or surface. Rapid decon may allow the unit to be in the lowest MOPP level possible and preserve its combat power.

PRINCIPLES OF DECONTAMINATION

The resources of manpower, time, and material are critical for your decision on how to sustain combat operations. Two concepts must be considered: the use of these resources and the ability to sustain combat operations. You must know when, where, what, and how to perform decon by following the four principles discussed in the following paragraphs.

First, decontaminate as soon as possible. This is the most important principle of the four. Consider this principle before you consider any other. Contamination hazards force you into higher levels of MOPP and immediately begin to degrade the ability to perform your mission. The sooner the contamination is removed, the sooner you can reduce MOPP levels.

Second, decontaminate only what is necessary. You cannot waste precious resources decontaminating everything. Decontaminate only what is necessary to continue your mission. Consider the following factors when you decide whether decon will interfere with the mission:

- Mission-"tempo of battle"
- Time available
- Degree of contamination
- Length of time you have been in MOPP 4
- Decon assets available

Third, decontaminate as far forward as possible (limit spread). Do not transport contaminated personnel and equipment away from your operational area if you can bring decon assets forward safely. This will keep your equipment on location where it is needed, allow decon to begin earlier, and limit the spread of contamination to other areas.

Fourth, decontaminate by priority. Clean important items of equipment first and the least important items last. The COC will prioritize the equipment for decontamination.

LEVELS OF DECONTAMINATION

Three levels of decon are used today: immediate, operational, and thorough. All three levels (fig. 6-1) are discussed in this chapter.

IMMEDIATE DECON

The aim of immediate decon is to minimize casualties, save lives, and limit the spread of contamination. Immediate decon is carried out by individuals upon becoming contaminated. Three immediate decon techniques are used today:

- Skin decon
- Personal wipe down
- Operator's spray down

OPERATIONAL DECON

The aim of operational decon is to sustain operations, reduce the contact hazard, and limit the spread of contamination. Operational decon will also eliminate the necessity or reduce the duration of wearing MOPP gear. Operational decon is done by individuals and/or units. It is restricted to specific parts of operationally essential equipment, material, and/or working areas. Operational decon minimizes contact and transfer hazards. Further decon may be required to reduce contamination to negligible risk levels. Two operational decon techniques are used today:

- Vehicle wash down
- MOPP gear exchange

THOROUGH DECON

The aim of thorough decon is to reduce or eliminate the need for individual protective clothing. Thorough decon is carried out to reduce

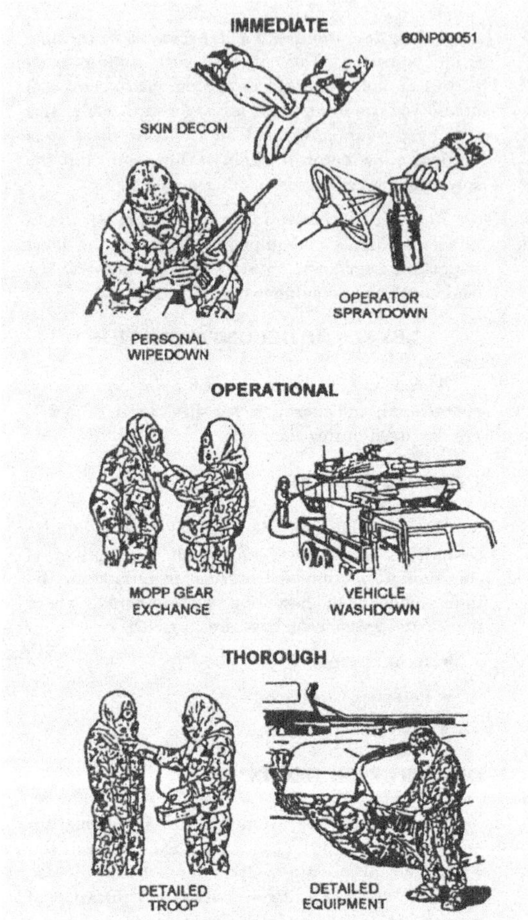

Figure 6-1.—Decon techniques.

contamination on personnel, equipment/material, and/or working areas to the lowest possible level (negligible risk). This permits the reduction or removal of individual protective equipment and maintains operations with the least degradation. Thorough decon reduces CBR contamination levels to a negligible risk. There are two thorough decon techniques that Seabees use:

- Detailed troop decon (DTD)
- Detailed equipment decon (DED)

We now know that contamination causes casualties and restricts the use of equipment and terrain. Decontamination reduces or eliminates the hazard and permits units to continue their mission. We also know the three levels of decontamination. We will now discuss decontamination operations for each level of decontamination.

IMMEDIATE DECON OPERATIONS

Immediate decon techniques are initiated by the individual Seabee, without command, once he or she is aware they have contamination on their bare skin. The individual Seabee uses his or her personal M291 or M258A1 skin decon kit to decontaminate exposed skin. Next, they decontaminate their MOPP gear and weapon, using the M280DKIE (decon kit, individual equipment) or with an additional M291/M258A1 kit. Use these items for chemical and biological contamination removal only.

Radiological contamination hazards affect you differently, but the principle is the same. Remove radiological contamination from equipment and personnel by brushing and/or using soap and water, respectively.

SKIN DECONTAMINATION

If chemical agents contact your skin, you must take immediate action to decontaminate yourself. Start the skin decon technique within 1 minute of becoming contaminated. Some toxic chemical agents, especially nerve agents, are rapidly absorbed by the skin and can kill in minutes.

Individual decon kits (IDK), M291 or M258A1 (fig. 6-2), provide the best means of skin decon. Instructions and procedures are on the outside of the kit.

The solutions in the M258A1 kit are caustic. Keep them out of the eyes, nose, wounds, and mouth. Use water to wash toxic agent or decon solution out of the eyes or wounds and seek medical treatment. Familiarize yourself with the operating instructions for both kits because of the differences in the M258A1 and M291 kits. The M258A1 uses a moist towelette with a decon solution, whereas the M291 uses a powder. Although you cannot see liquid contamination in the dark, you must be prepared to decontaminate during darkness. The M258A1 wipe 1 packet has an identifying tab that you can feel in the dark.

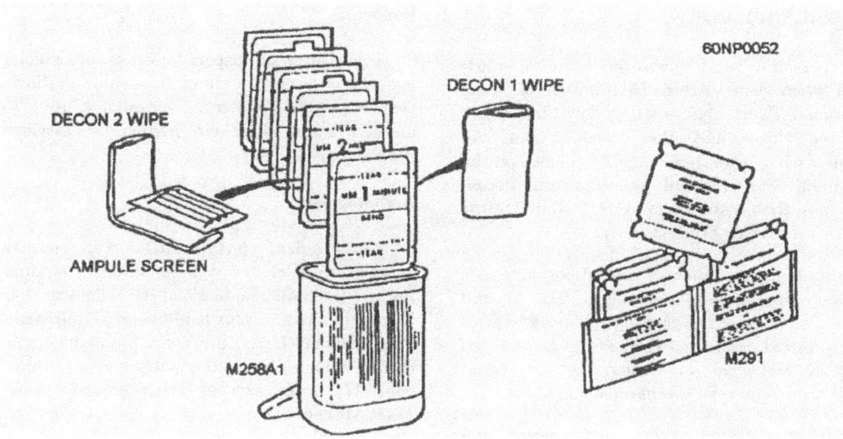

Figure 6-2.—Skin decontamination kits.

WARNING

Solutions in the M258A1 decon kit are flammable and unstable in storage at temperatures above 110°F (43°C) or for long periods in sunlight. Also, protect it from freezing temperatures. Used at 32°F (0°C) or below, the solution may cause frostbite.

If you do not have a skin decon kit, chemical contamination may be pinch-blotted from the skin with a cloth and the area flushed with water from your canteen. Pinch-blotting is better than rubbing because it limits the spread of the contamination. Soap, if available, can also be used to wash the agent from the skin. Washing with soap and water (or hot water) is the next best method for toxic agent decon. This method is not as effective as using the decon kits.

NOTE: The M291 skin decon kit replaces the M258A1 skin decon kit. However, there will still be a time when the two kits are available in the field.

Biological

Currently, no means exists of detecting biological agents. You probably will not know immediately when you have become contaminated. Most biological agents, except toxins, pose their primary threat through inhalation or ingestion. The skin is an effective barrier against most biological agents if it has no cuts or scratches.

Unit corpsmen know the types and levels of natural infection for the area of operations. They monitor these levels. If a given disease reaches a high level, they decide whether or not a deliberate biological attack has occurred.

The best biological defense is to take action before you are attacked. Keep immunizations up-to-date, observe basic sanitary precautions, and keep skin breaks covered. Treat minor cuts or abrasions by ordinary first-aid measures (iodine, Zephiran, or Merthiolate). Washing with soap and water removes nearly all biological agents from the skin. Frequent showering or bathing lessens chances of infection and disease. A 0.5 percent sodium hypochlorite (household bleach) solution is also an effective biological decontaminant. Appendix IV tells how to make this solution.

Radiological

Because no immediate life-threatening hazard is caused by radiological contamination, no immediate skin decon is required. However, wash exposed areas of your skin when possible. If your skin is contaminated by radiological contamination, use operational decon techniques immediately. (See "MOPP Gear Exchange" for a detailed discussion on reducing radiation levels from radiological contamination.)

PERSONAL WIPE DOWN

The personal wipe down technique is most effective when done within 15 minutes of being contaminated. Every Seabee wipes down his or her mask, hood, gloves, and other essential gear. (An exception is when a thickened agent is globed on the overgarment.) For chemical and biological decon, Seabees use their skin decon kits. Radiological contamination may be brushed away.

Do not attempt to remove chemical contamination from your protective overgarment. The special protective properties of the garment minimize hazards from chemical agents. However, brush off radiological, biological, or frozen chemical agent contamination from your overgarment.

If radiological contamination is not removed, your radiation exposure will increase over time.

Chemical

The stocks and handgrips of individual weapons also tend to absorb chemical agents. Once absorbed, they may present a vapor hazard for days. To reduce this penetration and vapor hazard, decontaminate individual equipment using the M291 or the M258A1 kits. You must decontaminate gloves, hood and mask, helmet, and weapon if they are contaminated. Perform personal and equipment wipe down within 15 minutes after being exposed to liquid contamination. Additionally, wearing your Kevlar helmet protective cover will prevent or reduce the absorption of any liquid chemical agent.

If an agent is globed on your overgarment, you may scrape it off with a stick or other object; otherwise, do not attempt to decontaminate chemical agents on your overgarment. This will provide little, if any, extra protection and you probably will not have enough M291 and M258A1 decon wipes to do so.

Decon wipe 1 works better against the G-type of nerve agents but can burn your skin. Decon wipe 2 works better against CX-type of nerve agents and blister agents. Additionally, decon wipe 2 is not as caustic to the skin and should help to neutralize some of the caustic compounds in wipe 1. When M291 and M258A1 decon wipes are not available, field expedient methods, such as washing with soap, water, and bleach solution, are partially effective (see appendix V).

Biological

If you know or suspect toxins or other biological agents are present, remove the contamination with soap and water. If water is not available, use M258A1 decon wipes in the same manner as described for chemical agent decon.

Radiological

Radiological contamination can readily be detected and located with monitoring equipment. Remove the contamination and you reduce the hazard. Brush the dust off your load-bearing equipment and mask carrier. If you are contaminated with a dry contaminant, such as fallout, shake your clothing and gear. Wash the exposed areas of your skin. Use M258A1 decon wipes if soap and water are not available. Pay particular attention to your hair and fingernails. Avoid breathing the dust you shake off by wearing a piece of cloth over your face. If you were contaminated by a wet radiological contaminant, you must immediately conduct a MOPP gear exchange. Brushing or shaking will not remove the wet radiological contaminant or its hazard. Wipe off your mask, hood, helmet, gloves, footwear covers, and other personal equipment with warm, soapy water. If warm, soapy water is unavailable, use rags or damp paper towels. Ensure contamination is not spread to clean areas.

OPERATOR'S SPRAY DOWN

After you have decontaminated yourself and your personal equipment, you may need to decon other mission-essential portions of your equipment before continuing your mission. For example, you may need to decon the vehicle you are operating or a crew-served weapon. To ensure you do not pickup contamination from these items, decontaminate those surfaces you or your crew must touch while operating the equipment. This decon is called the "operator's spray down" and is most effective when done within 15 minutes after personal wipe down. Starting this technique later is not as effective; contamination, especially chemical agents, will probably have spread and will be more difficult to remove by this technique.

Chemical

Decontaminate those surfaces you must touch to do your job. Use an onboard portable decon apparatus, such as M11 or M13 (fig. 6-3).

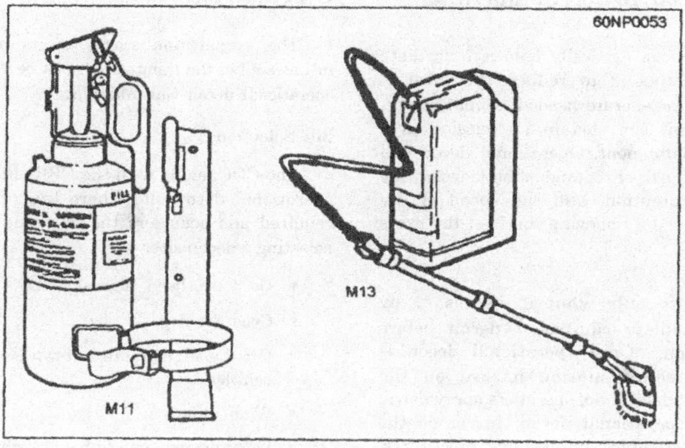

Figure 6-3.—Portable decon apparatus.

Spray DS2 onto those surfaces you must touch to do your mission. Scrub the DS2 into the surface with brushes. Wait 30 minutes, then wash it off. If a decon apparatus is not available, use the field expedient resources available to apply DS2 or supertropical bleach (STB) from the bulk containers. If necessary, use nonstandard decontaminants discussed in appendix IV.

Biological

A bleach solution is the preferred decontaminant for biological contamination; however, if it is not available, use hot, soapy water. Apply the bleach solution with brushes and scrub the surface well. Rinse the surface after scrubbing (a 30-minute wait is not required). DS2 or STB is also effective against most known biological contamination, but because of their caustic nature are not preferred. Other nonstandard biological decontaminants are described in appendix IV.

WARNING

Always use extreme caution when handling DS2.

Some important factors that should be remembered when handling DS2 are as follows:

- Do not mix DS2 and STB because mixing the two will cause a fire.
- DS2 is a combustible solution and should not be sprayed on personnel or protective clothing.
- Severe chemical burns can result if personnel fail to observe all safety precautions.
- DS2 can severely injure eyes and skin and, if inhaled, can cause illness.
- DS2 can damage CBR protective overgarments, protective gloves, hood, and overboots.

Radiological

If you are contaminated by fallout, rain out, neutron-induced contamination, or any type of radiological agent, use your monitoring equipment to help locate contamination. Then decontaminate as required. If detection equipment is not available and you suspect you are contaminated, decontaminate.

Radiological contamination can usually be removed by brushing or scraping (use brooms or tree branches). Water is effective for flushing away radiological contamination. Control the runoff by using drainage ditches that flow into a sump. Remember, you have not destroyed contamination, just moved it. The runoff will still be hazardous. Brushing or scooping away the top inch of soil from your fighting position will also lower the amount of radiological contamination.

OPERATIONAL DECON OPERATIONS

Operational decon generally follows immediate decon. The objective is to reduce the level of contamination to regenerate needed combat power. Therefore, the unit can sustain its mission in a contaminated environment. Operational decon will further reduce the risk of contamination transfer, the spread of contamination, and the speed of the weathering process by removing much of the gross contamination.

Decontaminate only what is necessary by conducting immediate equipment decon before operational decon. Once operational decon is completed, the contamination hazard on the equipment is neutralized. So, operator's spray down, combined with operational decon, increases the opportunities to conduct unmasking procedures. Operational decon is accomplished primarily by using the following assets currently in the battalion TOA:

- M12A1 or M-17 (contains pump, tank, personnel shower assembly, and water heater units)
- M11 (fire extinguisher-like device for spraying DS2)
- M13 (brush apparatus for scrubbing with DS2)
- Chemical agent monitor (CAM)
- Individual chemical agent detector (ICAD)

Vehicles must be identified as contaminated or noncontaminated before arriving at any operational decon station. If the contamination on the vehicle or equipment can be neutralized with immediate decon procedures, decontaminate and go on with the mission. To be most effective, you should accomplish operational decon as soon as possible.

MOPP gear exchange and vehicle wash down are done in conjunction. The COC establishes a MOPP gear exchange site, upwind of the vehicle wash area. This is the site where Seabees exchange contaminated MOPP gear for a reserve set of MOPP gear. The exchange is normally accomplished by squad-size elements. Unmasking may or may not be possible during this exchange. A well-practiced unit Standing Operating Procedure (SOP) will greatly simplify and ease the carrying out of operational decon procedures.

PREPARATION PHASE

The preparation stage, as its name implies, includes all of the things that must be done before any operational decon can take place.

Site Selection

The COC, along with the CBR officer, selects an operational decon site where little preparation is required and considers the following factors when selecting a decon site:

- Good overhead concealment.
- Good drainage.
- Off the main route but with easy access for vehicles.
- Wind direction.
- Large enough area to handle vehicle wash down and MOPP gear exchange for a squad-size element (100 square meters per site).
- A water source. Plan for about 100 to 150 gallons of water for each vehicle. Of course, larger or dirtier vehicles need more water.

NOTE: The M12A1 PDDA can carry 450 gallons to a decon site; the M17 LDS (with collapsible bladder) can hold 1,500 gallons but must be setup and filled up at the decon site.

Site Setup

The battalion decon crew will setup the vehicle washdown area. An operational decon site takes minimal setup and preparation. The site setup requires positioning the M12A1 or the M17 LDS power-driven decon equipment (PDDE) along the roadway, ready to dispense hot, soapy water. The vehicle washdown process consists of contaminated vehicles moving forward into a site to be washed down (removing gross contamination) and then moving out. Additionally, at least two Seabees that are contaminated set up a MOPP gear exchange site. Prepare MOPP gear exchange at a clean site 50 meters upwind of vehicle wash down. Both the vehicle wash down and MOPP gear exchange operations should operate concurrently.

If water for the M12A1 PDDE has been preheated, preparation for vehicle wash down should take no more than 10 minutes (this is a guideline). Preparation for vehicle wash down using the M17 LDS will take more time if the water bladder must be filled.

EXECUTION PHASE

This phase is the actual conduct of the two operational decon techniques: the vehicle wash down and the MOPP gear exchange. Companies will provide their own security while vehicle wash down and MOPP gear exchange are in progress.

When finished, troops mount their vehicles and move to their new battle positions. **For planning purposes, the vehicle washdown site will process one vehicle every 2 to 3 minutes; and MOPP gear exchange will take approximately 30 minutes.**

Vehicle Wash Down

Before vehicle washdown, units conduct the operator's spray down to increase decon effectiveness. Ensure that water pressure is from 60 to 120 psi; the pressure rate at which most gross contamination can be removed. Vehicle wash down is conducted as far forward as possible and is done by the battalion decon crew. The decon crew will use the power-driven decon equipment (PDDE) currently available in the TOA to conduct vehicle wash down, such as the M12A1 or M17 power-drive decon apparatus (PDDA). If the mission permits, it is most effective to conduct vehicle wash down between 1 to 6 hours after contamination. The longer you wait to remove or neutralize contamination, the harder it will be to do so. Also, the longer you wait, the more opportunity there will be for agents to spread and make contamination problems worse.

Each vehicle receives a 2- to 3-minute vehicle wash down with hot, soapy water from the PDDE. Vehicle operators maintain visual contact with each other to know when to move from concealment to the washdown area. The PDDE operator gives a signal when vehicles are required to move into position. Personnel in wheeled vehicles should dismount before wash down to avoid getting wet. Following this, the vehicle moves to the MOPP gear exchange area. Troops will dismount and conduct MOPP gear exchange by squad or vehicle.

Depending on the availability of equipment, METT-T, and the tactical situation, units have the option to select a one- (table 6-1 and fig. 6-4) or a two-lane wash down. When using two M12A1 or other combinations of decon equipment, two-lane operational decon speeds up the process (table 6-2 and fig. 6-5). Because speed is important and detection is difficult, do not check vehicle for contamination after vehicle wash down. Remove only gross contamination.

Mopp Gear Exchange

Mopp gear exchange is done by squads or crews so that leaders can control the rate of overgarment exchange. This procedure also ensures that adequate stocks of overgarments at company level are maintained and accounted for. Two Seabees can work as a buddy team, or a Seabee can execute the technique by himself. However, when squad leaders supervise, they can prevent unnecessary exchanges of MOPP gear by using the chemical agent monitor (CAM).

During the MOPP gear exchange, you change your contaminated MOPP gear for new, uncontaminated gear. The squad or platoon is responsible for conducting its own MOPP gear exchange at the assembly area of the operational decon site. Decontaminants and chemical suit replacements are provided by supply (S-4) and issued near the decon site.

MOPP gear exchange removes nearly all liquid or solid contamination from you and your individual equipment. When personnel have little, if any, vapor hazards on themselves, they may use hazard-free areas to unmask, to eat, to drink, and to rest temporarily. Before unmasking and lowering MOPP levels for temporary relief, conduct unmasking procedures using the CAM.

Eight steps are used in MOPP gear exchange. The squad forms a circle around a lead team, typically the squad leader and another Seabee. The troops are paired into buddy teams and spaced around the circle with 1 to 3 meters between teams. Members of each team alternate as they go through step 1 (decon gear). At steps 2 through 7, one member completes all steps with the assistance from the other team member. Roles are then reversed. See table 6-3 for detailed instructions on this method.

These techniques do not guarantee safe conditions required to allow unmasking on or near equipment. However, troops may move upwind of dirty vapor equipment into a clean area or collective protection shelter, check for contamination, and then briefly unmask. Conduct continuous contamination checks and monitoring to ensure you stay in clean areas. Use the M258A1 kit and check it every 15 to 20 minutes. Use the CAM every 5 minutes. Use the CAM together with unmasking procedures. Three types of procedures are used for MOPP gear exchange: buddy team, triple buddy system, and individual MOPP gear exchange.

Table 6-1.—One-Lane Wash Down

Steps and Risks	Equipment	Procedures
Step 1. Button up vehicle/equipment. Performing this step prevents contamination from being washed or splashed into uncontaminated areas. **RISKS.** Failure to perform this step may result in contamination being washed into uncontaminated areas, subjecting crew and maintenance personnel to hazards.	None	Equipment crew/operators close all access doors, hatches, windows, and other openings before washdown. Put muzzle covers on weapons. Nonessential personnel can dismount and begin MOPP gear exchange upon commander's orders. They then act as "buddies" for essential crew/operators.
Step 2. Washdown vehicle/equipment. Performing this step limits spread of contamination, minimizes hazard, and enhances weathering to make detailed equipment decon easier and faster. **Risks.** If you do not do this step, expect casualties from contact hazards. Spreading or transferring the hazard most likely will increase. Weathering of the hazard will be slowed. You will not be able to reduce MOPP level immediately because an after-vehicle washdown check for contamination is not made. (See "Decon in Combat," Chapter 1, for when to unmask for brief periods.)	Use for all forms of contamination: • One PDDE. • Adequate fuel for water heater (if available) and pump unit. • Adequate water supply (about 100 to 150 gallons per vehicle.) • Liquid detergent to mix with water (see Appendix J.)	**Chemical, biological, and radiological:** Two troops from the battalion PDDE crew operate washdown equipment. A third troop supervises. Troop must wear toxicological agent protective (TAP) aprons or wet weather gear worn over MOPP gear to keep MOPP gear from becoming saturated. Troops spray hot, soapy water (under pressure) from PDDE onto equipment surfaces. This removes, neutralizes, or destroys most of the gross contamination trapped in dirt mud. Unheated soapy water or plain water may be used, if necessary, but is less effective than hot, soapy water. Start at the top decks of vehicles and wash downward.
Step 3. Vehicles move into assembly area.		Exchange chemical suit.

SITE CLEARANCE PHASE

Although the operational decon operation is done rapidly with little site preparation, these areas will be contaminated when the operation is completed. This could be a hazard to friendly forces reoccupying the area.

Cleanup

The PDDE crew of the unit cleanup the MOPP gear exchange area. They bury or burn the contaminated refuse and retrieve any unused decontaminants. Burning will cause a downwind vapor hazard. **Burying is the preferred method of disposal of contaminated waste.** If you burn it, notify the COC; the COC will notify any units that may be affected by the vapor hazard.

The PDDE crew must control contamination runoff during the execution of operational decon. The PDDE crew should move the PDDE a few meters away from the vehicle washdown area and wash the decon equipment, including hoses, after the operation is completed. Wet weather gear or TAP aprons should be decontaminated with STB slurry and retained for future use. If MOPP gear exchange is done at a different location, the contaminated company will be required to clean up after itself.

Marking and Reporting

The PDDE crew of the battalion marks the operational decon site with standard NBC warning markers and reports the contaminated area to the COC. The COC sends out a NBC 5 report that alert friendly forces to avoid the area.

OPERATIONAL DECON SUMMARY

For operational and logistical purposes, units should plan to conduct vehicle wash down and

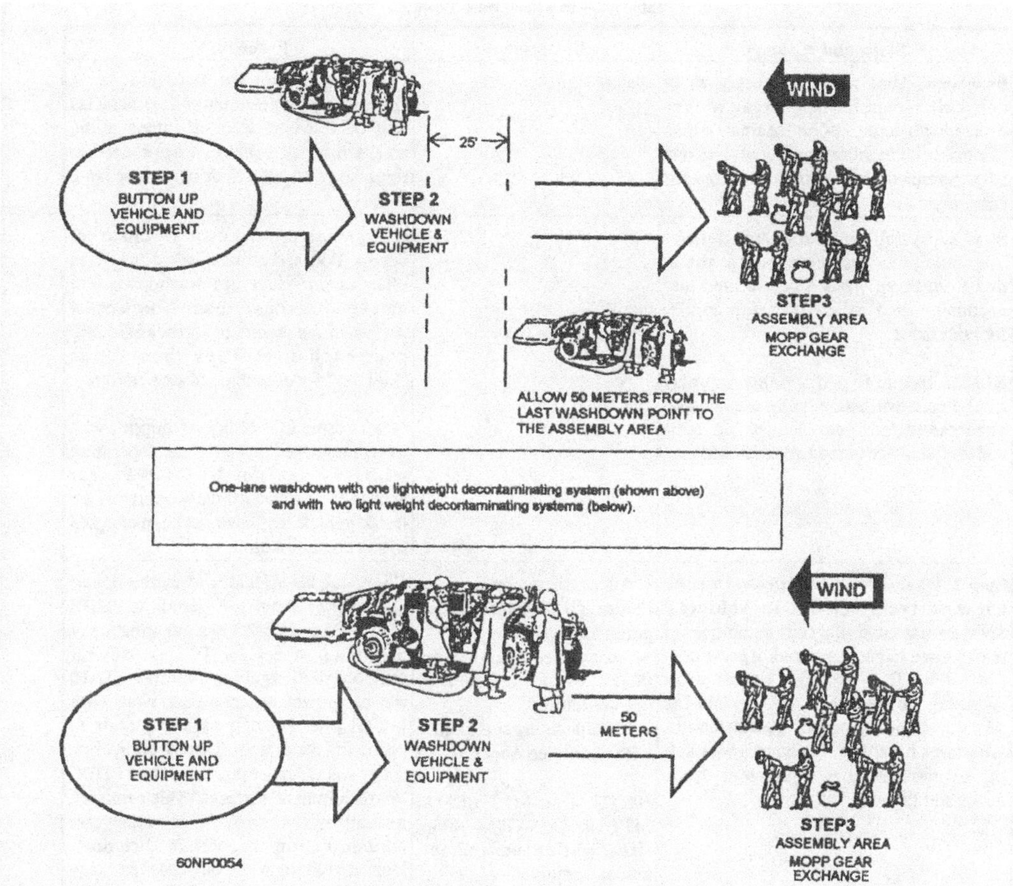

Figure 6-4.—One-lane wash down.

MOPP gear exchange concurrently. This should be done between 1 to 6 hours of becoming contaminated. This action reduces degradation and improves the ability of the unit to conduct its mission. Decontaminants and replacement MOPP gear are provided by the supply company with a vehicle near the decon site.

MOPP gear exchange and vehicle wash down are best used by squad-size or platoon-size elements. When larger elements try to process through an operational decon site, they lose many benefits of a small decentralized operation. Benefits of a squad- or platoon-sized decon operation include the following:

1. Tailored decon operations are flexible and responsive to small unit needs.

2. Small, speedy operations are more easily concealed in one location near the forward area.

3. A water source may not be needed at the decon site because most power-driven decon equipment (PDDE) have a water-carrying capability to support squad-sized elements.

Units must develop their own Standing Operating Procedure (SOP) for obtaining temporary relief from MOPP 4. The SOP is based on each unit's equipment and mission. Although every operation is unique, methods should be standardized when possible. In any case, personnel must know the following:

Table 6-2.—Two-Lane Wash Down

Steps and Risks	Equipment	Procedures
Preaction: Unit tactivally disperses in concealed marshalling area. Makes contact with control point for final orders. Control point monitors and supervises rate of movement into lanes to prevent congestion.	Watch	One soldier from the Battalion decon crew. Every three minutes two vehicles will be released from the marshalling area. On signal vehicles will proceed to the decon station in their respective lane.
Step 1. Button up vehicle/equipment. Performing this step prevents contamination from being washed or splashed into uncontaminated areas. This step applies for both lanes. **RISKS.** Failure to perform this step may result in contamination being washed into uncontaminated areas, subjecting crew and maintenance personnel to hazards.	None	Equipment crew/operators close all access doors, hatches, windows, and other openings before washdown. Put muzzle covers on weapons. Nonessential personnel can dismount and begin MOPP gear exchange. They then act as "buddies" for essential crew/operators. Note: Ensure that vehicles equipped with overpressurized systems are operating with systems on. No MOPP gear exchange is required if crew/operators are inside vehicle and have not been exposed to any contamination.
Step 2. Washdown vehicle/equipment. Crews/drivers remain in vehicles. Sprayers use cross diagonal technique for two to three minutes, removing gross contamination. This technique avoids water splashing the crew members. Performing this step limits spread of contamination, minimizes hazard, and enhances weathering to make detailed equipment decon easier and faster.	• Adequate fuel for water heater (if available) and pump unit. • Adequate water supply (about 100 to 150 gallons per vehicle.) • Liquid detergent to mix with water (see Appendix F.) NOTE: Use M12 PDDA, M17 LDS, 65-GPM pump, fire-fighting equipment, and/or combination.	**Chemical, biological, and radiological:** Two troops from the battalion PDDE crew operate washdown equipment. A third troop supervises. Troop must wear toxicological agent protective (TAP) aprons or wet weather gear worn over MOPP gear to keep MOPP gear from becoming saturated. Troops spray hot, soapy water (under pressure) from PDDE onto equipment surfaces. This removes, neutralizes, or destroys most of the gross contamination trapped in dirt mud. Unheated soapy water or plain water may be used, if necessary, but is less effective than hot, soapy water. Start at the top decks of vehicles and wash downward.
Step 3. Vehicles move into assembly area. MOPP gear exchange is determined by the commander.		Exchange chemical suit.

1. How to recognize and understand contanimtion hazards and how to avoid contamination when possible.

2. How to protect yourself and your equipment if contaminated.

3. Know the capabilities and limitations of MOPP gear.

4. Know how to neutralize or remove contamination hazards.

5. Do only as much decon as you need to continue your mission.

The following sample checklist (table 6-4) can be used by the battalion decon team for planning operational decon operations:

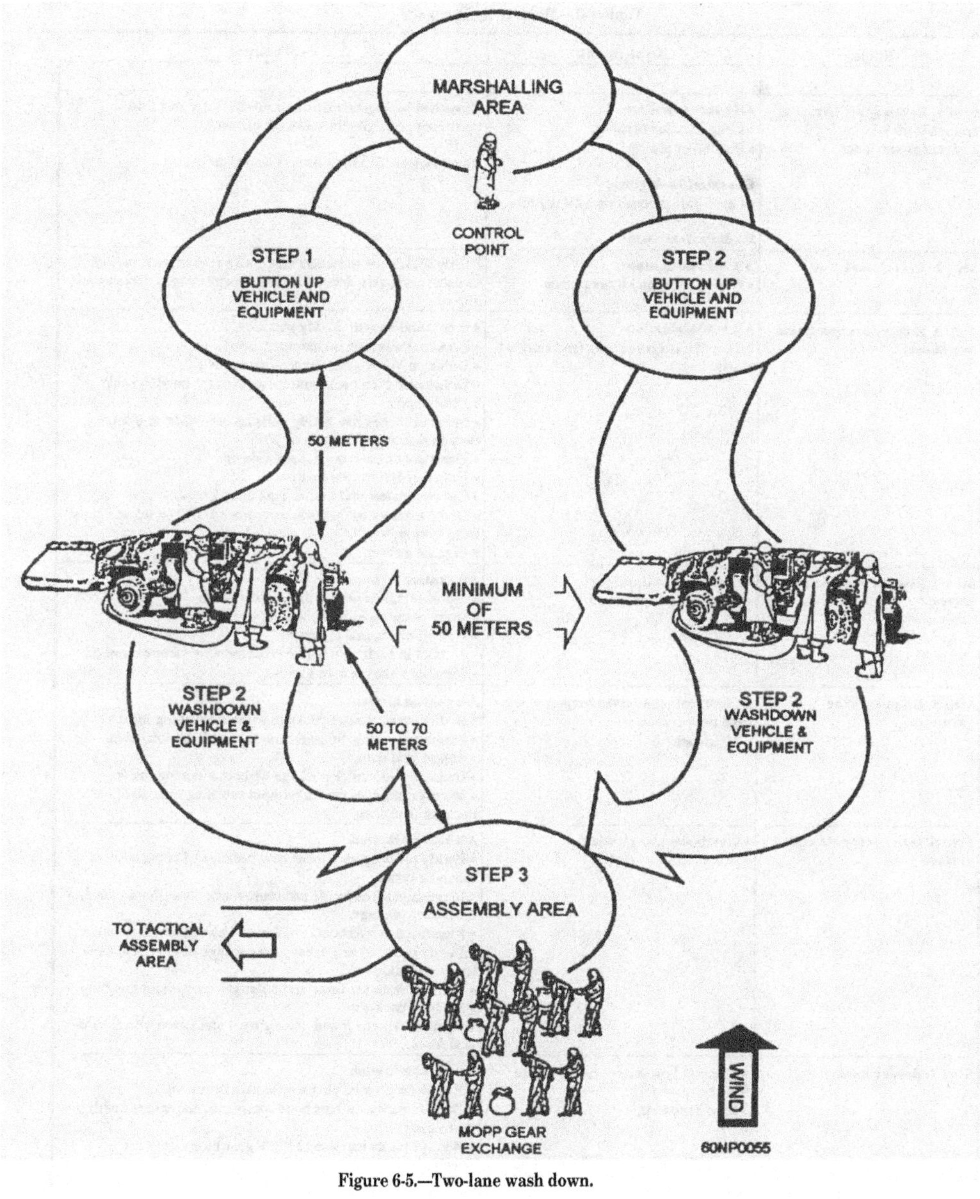

Figure 6-5.—Two-lane wash down.

Table 6-3.—Mopp Gear Exchange

Steps	Equipment	Procedures
Step 1. Decon gear: (weapon, helmet, load- bearing equipment, and mask)	**All contamination:** • Long-handled brushes. • Poncho or plastic. **Chemical/biological:** • One 5-gallon container STB dry mix. **Radiological:** None	**Chemical/biological:** Brush or rub STB dry mix into personnel gear. Gently shake off excess. **Radiological:** Shake or brush contamination off.
Step 2. Decon mask and gloves:	**All contamination:** • One skin decon kit per person	Buddy will decon members mask, all exposed parts including canister using skin decon kit. Each member then decons their gloves.
Step 3. Remove overgarments/ overshoes:	**All contamination:** • Two discard containers (recommend plastic bags).	**All contamination:** Buddy will: • Loosen drawstring on members hood. • Unfasten Velcro tabs on members smock. • Cut smock from back-bottom, up through hood at neck. • Lower hood. • Member makes fist, buddy pulls smock off from front, turning smock inside out. • Place smock on ground, back side up. • Unfasten Velcro tabs on trouser. • Cut suspenders and trouser legs up to knees. • Grasp trousers by cuff while member pulls one leg at a time from trouser. • Discard trouser.
Step 4. Remove overboots and gloves:	**All contamination:** • Containers from step 3.	**All contamination:** • Member stands next to discarded smock. • Untie or cut laces on boots. • Pull off boots, one at a time. • As boot in each boot is removed, member steps on smock. • Member removes own gloves.
Step 5. Don new over-garment:	• Chemical protective overgarment (one per person). • M9 paper.	**All contamination:** • Buddy opens trouser package without touching inside. • Member removes trousers without touching outside of package, and dons. • Buddy opens smock package without touching inside. • Member removes smock without touching outside of package, and dons.
Step 6. Don overboots and gloves:	• Overboots and gloves. • M9 paper.	**All contamination:** • Buddy picks up and opens new package of boots without touching inside. • Member reaches inside and removes boots without touching outside of package. • Member don overboots and fasten Velcro tabs on trousers. • Buddy picks up and opens new package of gloves without touching inside. • Member removes inner and outer gloves without touching outside of package. • Member don inner and outer gloves and fasten Velcro tabs at sleeves.
Step 7. Secure hood:	• Chemical protective overgarment (one per person). • Skin decon kit.	**All contamination:** • Buddy decon own gloves with skin decon kit. • Buddy raises members hood, positions, and secures with drawstring. • Buddy checks members MOPP gear thoroughly for openings.
Step 8. Reverse roles, repeat steps 2 through 7 and secure gear:	N/A	

Table 6-4.—Operational Decon Checklist.

1	Decon assessment	If the unit has no decon assets, COC will request assistance from higher headquarters.
2	Coordination	COC coordinates with the decon team of the unit. The decon should be done between 1 to 6 hours after becoming contaminated.
3	Site selection	COC selects the decon site and considers the following items: • Off main route but easy access • Large enough area (100 square meters per site for a squad-size element) • Good overhead concealment • Water source (100 gallons per vehicle) • Good drainage
4	Site setup	Ensure the PDDE is positioned properly and ready to dispense hot, soapy water. Ensure that each company operates the MOPP gear exchange simultaneously at the vehicle wash down.
5	Site control and security	Ensure vehicle operators know when to move into position at the washdown location. Ensure companies have provided site security.
6	Processing	Decon petty officer in charge (POIC) ensures vehicles are processed at a rate of 3 minutes per vehicle. Ensure personnel are conducting MOPP gear exchange, if needed.
7	Cleanup	Decon petty officer in charge (POIC) ensures the MOPP gear exchange area is cleaned up.
8	Marking and reporting	NBC officer ensures the decon site is properly marked and assists the COC in sending out an NBC 5 report to higher headquarters.

It is important to remember that performing operator's spray down before hasty decon will decrease the contamination transfer and increase the survivability of the crew.

THOROUGH DECON OPERATIONS

Thorough decon operations reduce contamination to negligible risk levels. They restore combat power by removing nearly all contamination from unit and individual equipment. Thorough decon enables troops to operate equipment safely for extended periods at reduced MOPP levels. A small risk remains, so periodic checks with the CAM, M8/M9 paper, or M256-series kit must be made after each operation. After thorough decon, the unit moves out of the decon site into a tactical assembly area. The unit, while in this tactical assembly, may undergo reconstitution or prepare for future operations.

Operational decon requires fewer resources but physically removes only surface contamination. Thorough decon is the most effective type of decon and is the most resource intensive. Weathering causes a significant reduction of contamination overtime, but the immediate reduction of contamination to negligible risk levels does not normally occur. Weather conditions, agent used, mission requirements, time, troops, and supplies available may all combine to make weathering the decon option of choice. Based on the recommendation of the CBR officer, the COC decides which type of decon is the most efficient and effective for their unit.

A limiting factor that must be considered when planning any decontamination operations is the availability of water. A typical vehicle will require 500 gallons of water during detailed equipment decon. In a water-scarce environment, such as deserts, the CBR officer must cordinate a water resupply plan with other units in the area of operation. A water resupply plan can include selecting a series of link-up points along a route. The unit can link up with a bulk water truck from other units. More complex water resupply plans include linking up with bulk water trucks, caching water throughout the area of operations, coordinating for the movement of water bladders by aircraft, and the identification of water sources in the area of operation of the unit. The use of nonpotable, salt, and brackish waters should be considered.

The thorough decon site consists of four main areas: the predecon staging area, the postdecon assembly area, the detailed troop decon (DTD), and the detailed equipment decon (DED). Selection of all four areas are based on operational guidance, road network, available cover and concealment, and water supply.

The predecon staging area is used by the contaminated unit to ready them for detailed equipment decon (DED) and detailed troop decon (DTD) operations. The process at the DED includes removing, neutralizing, or reducing contamination on interior and exterior surfaces of equipment to negligible risk. The DTD is the process of decontaminating individual fighting equipment to negligible risk and removing contaminated MOPP gear from the troops.

The postdecon assembly area is the location where vehicles and troops exiting the DED and DTD are linked. This is done before moving from the decon site.

PREDECON STAGING AREA

The contaminated unit, by company, moves to a predecon staging area approximately 250 to 500 meters downwind of the thorough decon site. Here, the contaminated unit conducts the required predecon actions that are designed to prepare the unit for the thorough decon operations. Predecon actions include the following:

1. Segregate contaminated vehicles from uncontaminated vehicles, if possible.
2. Allow vehicle crews to dismount and prevent further transfer/spread of contamination.
3. Prepare vehicles for detailed equipment decon.
4. Move contaminated vehicles and troops to the detailed troop and equipment decon lines.

Segregate Vehicles

Check all vehicles for contamination, using detection equipment. For chemical contamination, use the CAM and M8 detector paper. Also, visually check the vehicles for contamination using M9 paper. M9 paper affixed to vehicles reveals the presence of contamination and provides an indication of the level of contamination (fig. 6-6).

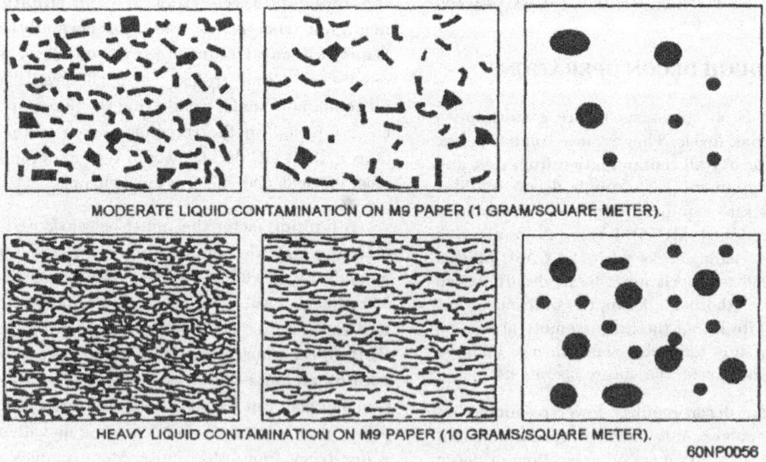

Figure 6-6.—Contamination levels.

When using the chemical agent monitor (CAM), such as the M256A1, ensure there is sufficient distance between vehicles. If the vehicles are parked too close together, vapor drift from contaminated vehicles may falsely identify uncontaminated vehicles as contaminated.

For radiological contamination, use the AN/PDR-27-series or AN/PDR-43 radiac detectors. If the vehicle only has isolated areas of contamination, use a portable decontaminating apparatus (M11) to decontaminate the chemically contaminated area. Vehicles found to have no contamination are sent to the reconstitution assembly area.

Crews Dismount

The vehicle crews, except drivers, dismount. As the crews dismount, they remove all equipment from the top of the vehicles. Once crew members have exited the vehicle, they will not reenter. This prevents further contamination from being spread into the vehicle interior.

Prepare Vehicles

The crews prepare their vehicles for processing through the detailed equipment decon (DED). All heavy mud and debris are removed from the vehicle by the crew. The crew should concentrate on the vehicle undercarriage. The reason for this is twofold: (1) contamination will most likely collect here and (2) the undercarriage is the hardest place to decontaminate. Tools used by the crew are placed back on the vehicle when finished.

Seat covers (when applicable), canvas items, camouflage netting, and any other materials that can absorb liquid contamination are removed from the vehicle. These items create a potential transfer hazard and are not easily decontaminated. Appendix V provides guidance on hard-to-decon items. Left untreated, absorbed chemical agents will desorb after being decontaminated and create a vapor hazard. Items that cannot be decontaminated by the standard methods used in detailed troop decon (DTD) are also removed and placed at a collection point.

Move To DED and DTD

In coordination with the decon petty officer in charge (POIC), the contaminated unit will begin to send contaminated vehicles and personnel to the DED and DTD. The COC has prioritized the vehicles for processing, sending the most important first. Communication is maintained between the predecon staging area and the POIC. All assistant vehicle drivers are the first individuals sent through the DTD to ensure there is a driver exchange at station 3 of the DED.

POSTDECON ASSEMBLY AREA

The company assembles in the postdecon assembly area after completing DTD and DED operations. The unit occupies this area before moving to a reconstitution location. The decon POIC will select the general location. It should be big enough to hold the entire unit undergoing thorough decontamination with proper cover and concealment. The assembly area should be located approximately 1 kilometer upwind from the DED and DTD.

DETAILED TROOP DECON (DTD)

Removing contaminated MOPP gear, including the protective mask, is the major action in detailed troop decon. If DTD is not performed, chemical agents may eventually penetrate the overgarment and contaminate underclothing or skin. How long the chemical agent will take to penetrate the clothing depends on the condition of the MOPP gear and the amount of agent on the gear. If contaminated with radiological contamination, the hazard will remain until removed.

The contaminated unit is responsible for setting up, operating, and closing the DTD in a thorough decon site. The CBR officer recommends to the COC the general location of the DTD within the decon site. He or she will also provide the COC technical advice on the setup, the operation, and the closure of the DTD.

The DTD has eight stations. The following pages discuss actions at each station and the resources required.

Station 1—Individual Gear Decon

Actions at this station remove contamination to a negligible risk level from individual equipment (782 gear, mask carrier, helmet, and weapon).

PREPARATION— The following equipment and supplies are recommended for station 1:

- Three 30-gallon containers
- Two long-handled brushes
- Sufficient supertropical bleach (STB) slurry mix

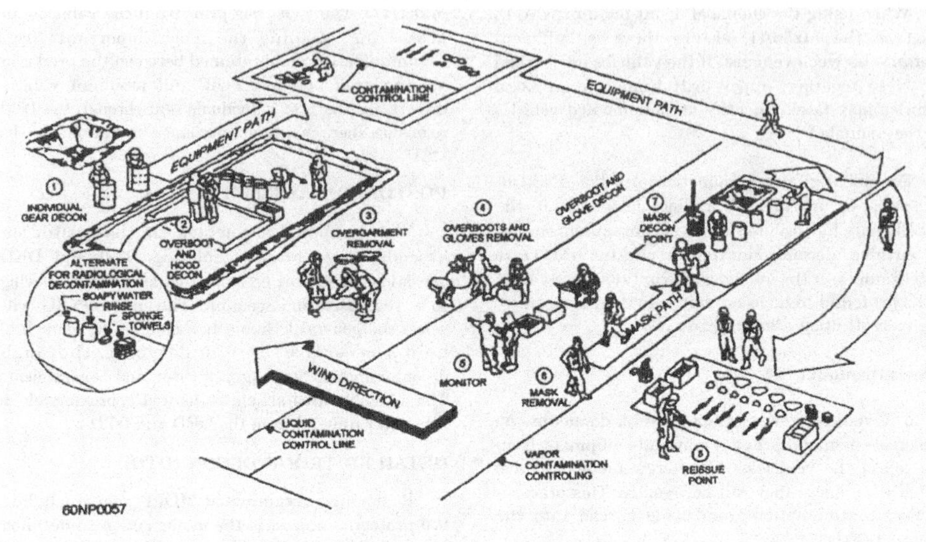

Figure 6-7.—Detailed troop decon layout.

- Two ponchos or plastic tarps

Dig a sump 6 feet square and 4 feet deep. Place three 30-gallon cans near the sump (fig. 6-7).

Fill two cans with an STB slurry mixture. The other can is filled with clean water for the rinse. Place two long-handled brushes at each can of the STB slurry.

To prepare the slurry, mix 100 pounds of STB with 20 gallons of hot water. Change the mixture after 20 troops have decontaminated their gear. Change the rinse water after every 10 troops or when it appears dirty.

Place a poncho or plastic tarpon the ground at the checkpoint. Divide the poncho or tarp in half, using tape. **This is the contamination control line (CCL).** The checkpoint will beat a minimum of three meters from all other stations. This will ensure a true reading on the detection equipment. Place the following chemical detection/identification equipment at the check station for an average company-sized unit:

- One CAM
- Eight books of M8 chemical detector paper
- Four M256-series chemical agent detector kits
- One hundred plastic trash bags

Three troops are required to operate this station. One supervises the decontamination of the individual gear and takes the decontaminated equipment to the check station. He or she also prepares new slurry mixtures of STB as necessary. One troop remains at the checkpoint and checks all gear for completeness of decontamination, using the detection equipment. One troop transports the decontaminated gear to the reissue station.

EXECUTION— Decontaminate the gloves with the water or STB slurry mix. If wearing the M24, M25A1, or M42 mask, use hot, soapy water and sponge or STB dry mix to decontaminate the hose and canister. Decontaminate the gear by washing it in the decontaminant container and scrubbing the gear for 6 minutes. Then dip the gear into the rinse water for 4 minutes. The station attendant will take the gear to the equipment checkpoint. **Place the decontaminated gear on the "dirty" side of the contamination control line (CCL).** The monitor checks the gear for contamination using the appropriate detection devices. The monitor holds the detection device 1 inch from the gear and checks for residual contamination. If the residual contamination exceeds negligible risks, recycle the gear and decontaminate it again. If the gear passes the check, place it **on the clean side** of the contamination control line. An attendant carries the equipment to the reissue point.

You must understand that the preceding paragraph outlines the minimum that should be done. Depending on time available, more extensive washing and checking procedures may be useful. You may decide to wash equipment longer or let it air outside after washing. This further reduces the amount of agent vapor desorbing (or bleeding) out of the pistol belt, mask carriers, helmet covers, and so forth. Equipment can be put in closed areas or plastic bags and checked for hazardous vapors with CAM or M256-series chemical agent detector kit. The CAM will only detect the nerve agent (G) and blister agent vapors (H).

Station 2—Overboot and Hood Decon

Actions at this station neutralize gross contamination on overboots and lower trouser legs. Gross contamination on the exposed parts of the protective mask is also decontaminated.

PREPARATION— Prepare a shuffle pit by digging a shallow pit about 3 feet wide by 3 feet long by 6 inches deep. Fill the shuffle pit with a supertropical bleach (STB) dry mix. Prepare the STB dry mix by mixing 3 parts earth to 2 parts STB. Add more STB to the mix after ten people have processed through the shuffle pit. The following equipment/supplies are needed at this station:

- One M258A1 or M291 skin decontaminating kit per person
- 6 feet by 6 feet piece of plastic
- Trash bags (as required)
- Ten drums of STB

Only one decon team member is required to operate this station. The buddy system is used to decontaminate the overboots and protective mask.

EXECUTION— Walk into the shuffle pit that is composed of STB dry mix (3 parts earth and 2 parts STB). Use gloved/protected hands to rub STB dry mix on the boots and lower trousers. Decontaminate the hood according to one of the two following procedures. Use the skin decon kit to decontaminate the hood and exposed parts of your buddy's mask. When you have finished decontaminating your buddy's mask, decontaminate your gloves with the skin decon kit in preparation for decontaminating your buddy's mask.

NOTE: Buddies should check each other's overboots and overgarment for damage. Any rips, tears, or punctures in overboots, rubber gloves, or overgarment should be reported to the attendant at station 5. This will allow the attendant at station 5 to check you for chemical agent symptoms and your clothing for possible contamination.

Station 3—Overgarment Removal

Performing this step ensures removal of the contaminated overgarments before an agent penetrates the overgarment material and touches the undergarments or skin.

PREPARATION— The materials and equipment needed at this station are as follows:

- Cutting tool
- Two 30-gallon garbage cans
- One hundred plastic bags (or at least one per Seabee)
- Ten boxes M258A1 or M291 decon kits

This station requires only one decon member. He or she supervises the actions of the troops processing through the station.

EXECUTION— Buddy teams will divide and continue the decontamination process individually. The station attendant assists in the removal of the smock and trouser by cutting them off.

CAUTION

Ensure that contaminated parts of the trousers and smock do not fold in and touch the clean uniform.

The station attendant cautions the troops not to rub the contaminated boot against his or her clean leg and to step wide enough so as not to rub the clean leg against the overgarment. If this step is done improperly, the agent may be transferred to the underclothing or skin.

Station 4—Overboots and Gloves Removal

This step removes contaminated overboots (footwear covers) and gloves to limit the spread of contamination. Overboots and gloves are also decontaminated for reissue.

PREPARATION— The following equipment and materials are needed:

- Tape
- Cutting tool
- Eight 30-gallon containers (garbage cans)
- One hundred plastic bags (or one per Seabee)
- Two scrub brushes
- Two ponchos or large plastic sheets
- CAM
- M8 paper
- 10 percent STB/HTH solution
- Hot, soapy water
- Cold rinse water

The station will be setup as shown in figure 6-7. Fill two 30-gallon containers with hot, soapy water and two containers with 10 percent STB/HTH solution. Place the two scrub brushes near the containers. Fill two 30-gallon containers with cold rinse water.

Two decon team members are needed to decontaminate the overboots and gloves. One person processes the overboots, while the other person processes the gloves.

Use the tape to mark the liquid contamination control line (LCL) on the ground clearly. The LCL separates the "dirty" from the "clean" areas. Place the cutting tool, two containers, and the plastic bags on the "dirty" side of the liquid contamination control line.

NOTE: An exception to the clean area is the contaminated gloves. However, you must ensure no liquid agent is spilled on the ground beyond the liquid contamination line.

EXECUTION— The attendant unfastens or cuts the strings or elastic closures on the boots. The station attendant holds the boot while the Seabee steps out and across the control line. The action is repeated for the other boot. Boots are then discarded in the designated container by the station attendant. If the boot cannot be removed by this process, the attendant cuts off the boot. The Seabee works the chemical protective gloves loose using the pinch-pull method, and the station attendant pulls the gloves off. The station attendant discards the gloves in a container **on the "dirty" side of the control line.** Station 4 attendant does his or her duty from the dirty side of the liquid contamination control line.

The boots and gloves are decontaminated in the following process:

Step 1. Submerge the gloves and boots into the container of hot, soapy water. Some contamination will be removed during this step. When the boots and gloves are removed from the container, ensure that no water remains inside the boots and gloves. **Do not decon any item that is unserviceable.**

Step 2. Submerge the boots and gloves into the container of HTH solution. Thoroughly scrub the items until no visible contamination remains. After scrubbing, submerge each item again before moving them into the rinse container.

Step 3. Thoroughly rinse the scrubbed items, making sure that the items are rinsed inside as well as outside. Check all the gloves and boots for holes, tears, or punctures and discard any that are found.

Step 4. Dry the usable items. Discard any glove or boot having punctures, tears, rips, holes, or other damage into a pile. Place the usable items on the poncho or plastic sheet to air-dry and weather.

Step 5. Place the items into the plastic trash bags, along with an M256 detector ticket. If the detector ticket shows no contamination, the items can be reissued or stored for issue later. If the detector ticket shows contamination remaining, the station attendants can recycle the items or discard them.

Obviously, you cannot check for the agent on the combat boots if you do not remove the overboots. If the overboots are not properly removed, you risk contaminating your combat boots and spreading contamination to clean areas. If rubber gloves are not properly removed, you risk contaminating your skin and underclothing at this and the following stations.

Station 5—Monitor

Performing this step identifies contamination on personnel, provides spot decon capability, and provides medical aid, as required.

PREPARATION— The following materials and equipment are needed:

- First-aid supplies
- CAM
- Five packs M8 detector paper per 100 people
- One case M258A1 or M291 skin decontaminating kits per person

Two troops are needed to operate this station. A corpsmen should be present to treat any personnel suffering from chemical agents symptoms.

EXECUTION— The monitor checks the individuals for contamination. Corpsmen will administer first aid as required. Liquid agent can be detected with the M8 detector paper. Small quantities of agent vapor can be detected with the CAM. Symptoms of agent poisoning are the most obvious indication of skin contamination. At this step, the medic checks each troop for symptoms and treats them as required. Troops should report any damage to their MOPP gear that was identified at stations 2, 3, and 4. Any areas identified as contaminated can be decontaminated with the M258A1 or M291 decontaminating kit by the station attendants. Individuals will be monitored after decon. It is possible that all liquid chemical contamination is absorbed into the clothing. Chemical detector paper (M8 and M9) will indicate negative, even though there is a hazard. The COC can choose to conduct more extensive contamination checks here if time is available. This will decrease the risk of casualties.

Station 6—Mask Removal

Performing this step removes the mask without contaminating the Seabee. The mask is taken to a mask decon point, limiting agent transfer at the site.

PREPARATION— The following equipment is needed:

- M8A1 automatic chemical agent alarm

Two troops are needed to operate this station. They remove and carry masks to the mask decon point (station 7).

EXECUTION— If the hood is still attached to the mask, the attendant pulls the hood over the front of the mask, grabs the mask by the voicemitter cover, and pulls the mask off the individual. The individual holds his or her breath as the mask is removed. If the mask has optical inserts, the attendant holds the mask open so the individual can remove the inserts without touching the outside of the mask. The individual then walks upwind 5 meters, crosses the vapor contamination control line, and resumes breathing. The station attendant brings the mask to station 7, mask decon point.

No chemical vapor hazard is expected beyond the vapor hazard contamination control line if the wind direction remains constant. Position the M8A1 or the M-21 automatic chemical agent alarm upwind of the site to warn of vapor hazards.

Station 7—Mask Decon Point

Performing this step removes all the contamination from the mask.

PREPARATION— The following material and equipment are needed:

- Four containers (about 3-gallon capacity)
- CAM (for chemical only)
- Hot, soapy water
- Rinse water
- Mask sanitizing solution
- Immersion heater 30-gallon container
- Two sponges
- One case of paper towels

Dig a sump (4 feet wide by 4 feet long by 4 feet deep) in which to discard used filters and canisters.

Three troops are needed to operate this station. Two troops will strip, wash, rinse, sanitize, and dry masks. The other troop checks the masks and carries them to the reissue point.

EXECUTION— Remove the eye lens, the outserts, and the hood, if the hood was not cut off at station 2. Remove and discard the filters or canisters. Put the items into the properly marked containers. Wash the mask, the hood, and the outserts in hot, soapy water. Rinse them in clean water, dip into sanitizing solution, agitate for 5 minutes, then rinse in clean water again. Add one tube of mask sanitizing solution (calcium hypochlorite) to each quart of water. Wipe with rags until they are almost dry. Discard each gallon of mask sanitizing solution into a sump after ten masks. The attendant checks for contamination with a CAM. If the mask is still contaminated the attendant recycles it for more decon and then decontaminates his or her rubber gloves. If the mask is not contaminated, the attendant takes the unassembled mask to the reissue point. The attendant must take care not to contaminate the reissue point or themselves.

Not properly performing this step will cause the risk of contaminating troops when reissuing masks at the reissue point. Even though the step is done correctly, there is still a possible danger when many masks are stacked together. Small amounts of residual vapor from each mask can become potentially dangerous.

Station 8—Reissue Point

Performing this step will ensure that you receive all of your individual gear and your protective mask with all the components.

PREPARATION— You will need the protective mask PLL parts. The unit supply company and chemical POIC set up the reissue point to provide the troops with replacement parts and assist in mask maintenance.

EXECUTION— Reissue the mask with the components to the troops who assembles the mask in the assembly area. Individuals pick up individual gear and move it to postdecon assembly area.

DETAILED TROOP DECON SUMMARY

Three lines are established when setting up for detailed troop decon: contamination control line (CCL), liquid contamination line (LCL), and vapor contamination line (VCL). It is critical that contamination is maintained at each line and does not cross each line. Eight stations are used for detailed troop decon. Table 6-5 is a company level detailed troop personnel and equipment recapitulation for each station.

Table 6-5.—Detailed Troop Decon Personnel and Equipment Recapitulation

Station	Personnel	Equipment
Station 1 Individual	2 decon team members 1 monitor (CAM operator)	3 30-gal containers 2 long-handle brushes 2 ponchos or plastic sheets 1 CAM 8 M8 detector paper 4 M256A1 kits 100 trash bags
Station 2 Overboot and Hood Decon	1 decon team member	60 M258A1 (one per person) 2 ponchos or plastic 100 trash bags
Station 3 Overgarment Removal	1 decon team member	2 cutting tools 10 M258A1 kits 2 30-gal containers 100 trash bags
Station 4 Overboot and Glove Removal	1 decon team member	2 30-gal containers 100 trash bags Tape Cutting Tool
Station 5 Monitor	1 CAM operator 1 Corpsmen	1 CAM 5 M8 detector paper kits 24 M258A1 kits
Station 6 Mask Removal	2 decon team members	1 M8A1 or M-21 chemical alarm
Station 7 Mask Decon Point	2 decon team members 1 monitor	4 3-gal containers 1 CAM 2 sponges 1 case paper towels 1 immersion heater w/container Mask sanitizing solution
Station 8 Reissue Point	Supply company POIC Decon POIC	Mask PLL

DETAILED EQUIPMENT DECON (DED)

ALFA Company is normally responsible for the setup, the operation, and the closure of the DED portion of the thorough decon operation. The COC will select the DED site. The DED for chemical and biological contamination is comprised of five stations. For radiological contamination, the DED uses all but station 2, DS Application. Actions at each station are described below. Appendix VI is a work/rest table for DED.

Station 1—Initial Wash

The objective of this station is to remove the gross contamination and dirt from the vehicle. The vehicle is sprayed for 2 to 3 minutes with hot, soapy water. The vehicle is then scrubbed to help remove caked-on dirt. The mechanical action of scrubbing also helps remove thickened chemical agents. Although the undersurface is difficult to reach, try to remove as much dirt as possible. This station will use approximately 250 gallons of water per vehicle. Larger vehicles or vehicles with large quantities of dirt will use more water. **The runoff from this station is contaminated and must be treated as hazardous.** This station requires high water pressure systems (M12A1 PDDA), rather than high water volume systems (65-gpm pumps).

Hot, soapy water is water heated to about 120°F to 140°F to which a detergent has been added. Hot water alone is less effective than hot, soapy water. Because of the high temperature, some agents are best removed by steam through vaporization. Finally, for some chemical agents cold water exhibits better solvent characteristics.

Station 2—DS2 Application

The objective of this station is to apply decontaminant to the entire vehicle. The vehicle is divided into four parts, and a member of the scrubbing team is assigned each part of the vehicle. This limits the work load of each member of the scrubbing team and avoids duplication of work. DS2 is applied starting at the top of the vehicle and working toward the undercarriage. Every effort is made to apply DS2 to the undercarriage, especially if the vehicle has crossed a contaminated area. The mop is the least tiring method of applying DS2. Using a mop to apply DS2 creates a large amount of spillage. However, continual use of the M13 (decon scrub brush) requires the scrub team to exert more energy than using the mop. In hard-to-reach places, use the M13 to apply DS2.

Before the start of this operation, the scrub team pours 5-gallon cans of DS2 into 30-gallon trash cans if mops are going to be used instead of the M13. Each member of the scrub team will wear wet weather gear to protect themselves from the DS2.

NOTE: Water adversely affects DS2's ability to react with chemical agents. When water in DS2 reaches 20 percent by weight, the reaction between DS2 and the chemical agent stops.

If there is excess water remaining on the vehicle from station 1, there are several options:

- Wait for most of the water to evaporate.
- Remove the excess water.
- Increase the amount of DS2 applied.

There must be sufficient DS2 on the item being decontaminated for complete neutralization to occur. The DS2-to-agent ratio needs to be 55 to 1 for H agents and 25 to 1 for G agents. DS2 should be applied with scrubbing. Scrubbing increases the mixing of the agent with DS2, especially when thickened chemicial agents are present.

Station 3—Wait/Interior Decon

The objective of this station is to allow the DS2 to neutralize the chemical agent and to decontaminate the interior of the vehicle completely. Vehicles are moved to a concealed position. Vehicles will remain in station 3 for **no less than 30 minutes.** When you allow the DS2 to remain on the contaminated surface for 30 minutes, the amount of agent that will later desorb (off gas) will be significantly reduced. When there is a 30-minute contact time, there will be no resorption after decon operations for most chemical agents. However, studies suggest that HD vapor will desorb after decon, even if DS2 is allowed to remain for 30 minutes.

While the vehicle is held in this station for the DS2 to react completely, the driver inspects the interior of the vehicle for liquid contamination. The driver will be given M8 detector paper. If the driver identifies chemical contamination, he or she will be given decon supplies to decontaminate the interior of his or her vehicle. The best decon solution for use in the interior of vehicles is a 5 percent solution of HTH or STB. The driver wipes all reasonable accessible surfaces with a rag or sponge soaked in the HTH or STB solution. Do not attempt to decon areas where there is little likelihood of contamination (electrical assemblies, etc.).

CAUTION

Do not mix HTH or STB with DS2. If mixed, a violent reaction will occur.

Once interior decon is completed, the driver dismounts from the vehicle and proceeds to the start of the detailed troop decon (DTD). Having completed the DTD, the assistant driver mounts the vehicle and moves it to the next station. Drivers must exercise caution when entering or exiting the vehicle. A DS2-coated surface is slippery and DS2 may react with chemical protective footwear. Personnel should avoid stepping in DS2 and tracking it into the vehicle.

For radiological contamination, use an AN/PDR 27-series radiac meter to decide the extent and location of contamination inside the vehicle. If there is contamination, figure out the intensity of the contamination inside of the vehicle. If the contamination has an **intensity greater than 0.33 cGy (the negligible risk),** the interior of the vehicle must be decontaminated. Use hot, soapy water to wash the contaminated areas. Use a sponge to mop up the water and the contamination. Table 6-6 is a planning guide for decon mixtures.

Station 4—Rinse

The objective of this station is to remove the DS2 from the vehicle. The vehicle is sprayed with water from top to bottom. Take care to rinse the undercarriage. This station uses approximately 200 gallons of water. Failure to remove all the DS2 from the vehicle may cause false positive readings at station 5. If high water pressure systems (M12A1 PDDA, M17 LDS) are not available, large volume water pumps (65-gpm pumps) should be used at this station.

Station 5—Check

The objective of this station is to check the completeness of the decon. This station determines whether the vehicle has a negligible risk or still has significant contamination remaining. Detection procedures will vary depending on the type of contamination. If significant contamination is found on the vehicle, the vehicle will be recycled to station 2 for chemical contamination or station 1 for radiological contamination.

Chemical

The CAM is used to check for the presence of vapor from residual liquid contamination. A one bar or lower reading on the CAM indicates a negligible risk. Once the CAM indicates the presence of vapor contamination, M8 detector paper is used to verify the presence of liquid contamination. If it is suspected that both the CAM and M8 paper are producing a false positive reading, use an M256 chemical detector ticket to confirm or deny the presence of contamination. Table 6-7 is a list of common interference that can cause false positive readings on the CAM. If the vehicle has significant contamination remaining, recycle it.

There will be desorption of chemical agents from the surfaces after decontamination. On CARC-painted surfaces, the desorption of vapors will stop sooner than alkyd-painted surfaces. Consider this when checking decontaminated items for overall decontamination effectiveness.

Table 6-6.—Preparation of Decon Mixtures

Decon Mixtures		
Solution	**1 gallon of water**	**5 gallons of water**
5 percent	.6 pounds STB/HTH	3.6 pounds STB/HTH
10 percent	.75 pounds STB/HTH	4.5 pounds STB/HTH

Table 6-7.—Common Interferences That Can Cause False Positive Readings on the CAM.

Interferant	G-Bar Response	H-Bar Response
M258A1 decon kit		High
M280 DKIE		High
DS2	Low	
Insect repellent	Low-Very High	
Brake fluid	High-Very High	Very High
Cleaner, general purpose	High	
Burning kerosene		High
Breath mints	High	
Gasoline vapor	Low	Low
Burning grass	Low-High	Low
Burning gas	Low	
Green smoke	Low	Low-High
Break-free oil	Low	
Ammonia	Very High	

Radiological

Use the AN/PDR-27-series to decide if any contamination remains. If there is contamination remaining, determine the intensity of the contamination inside and outside of the vehicle. If the contamination has an intensity greater that 0.33 cGy (the negligible risk), the vehicle is recycled to station 1.

Recycle Criteria

The COC, together with the CBR officer, will establish the recycle criteria before the start of decon operations. The recycle criteria decides which vehicles return to station 1 after contamination is detected at station 5. If the unit has sufficient time and resources, any vehicle having more contamination than a negligible risk should be recycled. However, time and resources are usually limited and not all vehicles can be recycled. The recycle criteria is based on the weathering effects.

DED Configurations

Decon teams, establishing thorough decon sites, may vary because of organizational and equipment differences. This section describes the optimum setup configuration. The optimum configuration provides the maximum output for decon teams at 100 percent personnel and equipment. The equipment and personnel requirements for the optimum configuration are identified for both the decon team and supported company augments.

ALTERNATE LAYOUT PLANNING CONSIDERATIONS.— The CBR officer uses METT-T to determine the best possible DED layout for executing the mission. When determining alternate DED layouts, you should apply the following guidelines:

- Retain the ability to spay hot, soapy water or steam under pressure at station 1.
- Station 2 requires the largest number of persons.
- Experienced and qualified CAM operators are required at station 5.
- Water does not have to be hot to rise off DS2; however, the lower the water pressure, the more water required for the rinse.

DECON TEAM EQUIPPED WITH AN M12A1 PDDA.— The optimum setup of a DED for a M12A1 PDDA-equipped decon platoon requires the use of all authorized equipment and personnel (fig. 6-8).

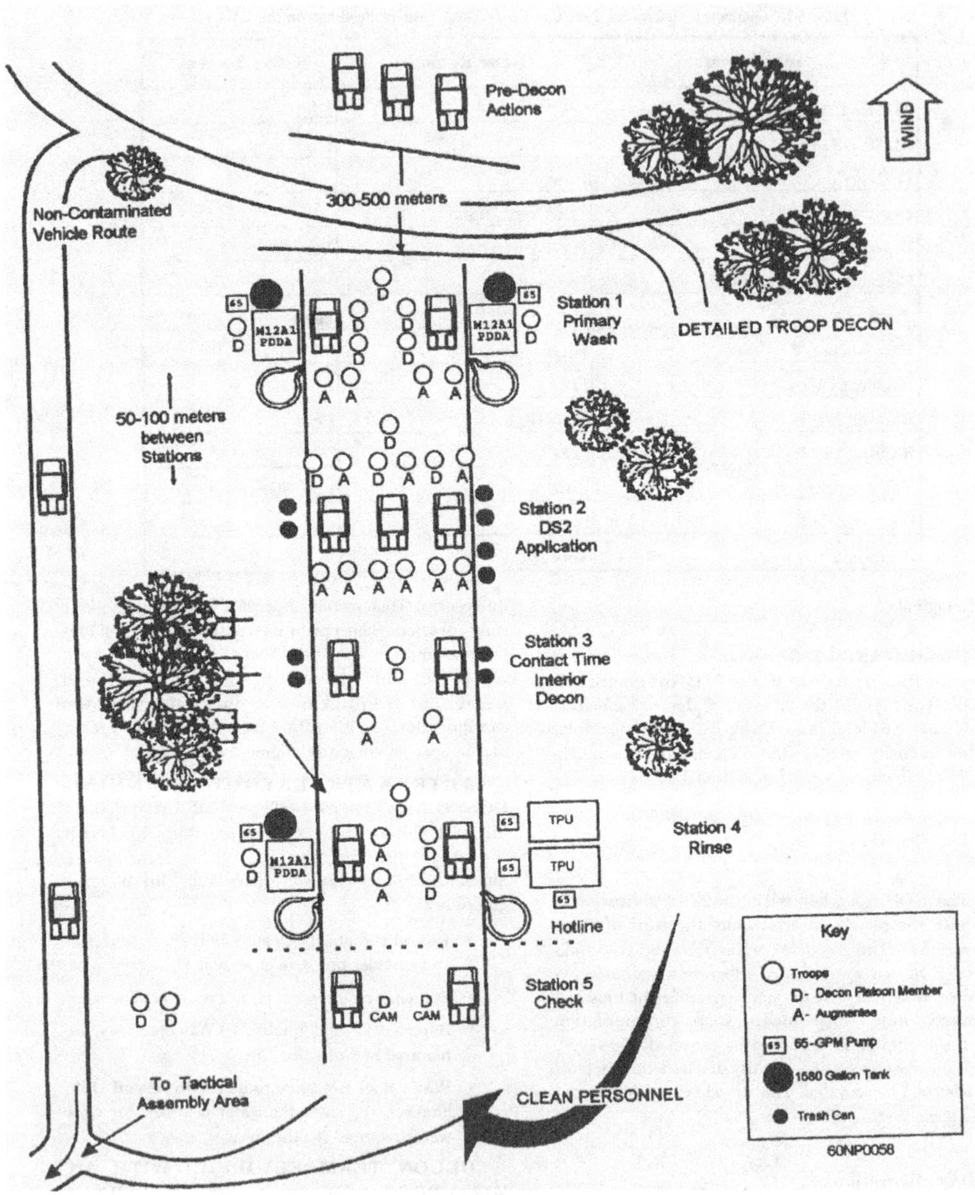

Figure 6-8.—Optimum DED for an M12A1 PDDA-equipped layout.

While this DED configuration is manpower and equipment intensive (table 6-8), it provides for the rapid decontamination of vehicles and equipment (eight vehicles processed per hour).

This layout uses dual lanes at stations 1, 4, and 5 to process two vehicles at once. Since the most time and labor-intensive work takes place at station the 3, this station is designed to process three vehicles at a time. The processing rate of this configuration mission will be affected by any work/rest cycle

WARNING

Work/rest tables are found in appendix VI.Chemical unit leaders must consider the impact of the work/rest cycle on the ability of their operation to process vehicles through DED. Failure to initiate a work/rest cycle could result in heat casualties and failure.

Table 6-8.—Optimum M12A1 PDDA-Equipped DED Setup

Station	Decon Personnel	Augments	Equipment
Station 1 Initial Wash	1 sqd l dr 2 PDA op 4 sprayers	4 scrubbers	2 M12A1 PDDAs 2 3,000-gal tanks 2 65-gpm pumps 6 long-handled brushes 8 TAP aprons Liquid detergent
Station 2 DS2 Application	1 sqd l dr 3 appliers	9 appliers	18 long-handled brushes 9 mops w/extra mop heads 3 30-gal containers 9 M13 DAPs Sufficient DS2
Station 3 Wait/Interior/Decon	1 POIC	2 assistants	2 AN/PDR 27 3 TAP aprons 6 30-gal containers 10 M8 detector paper 30 sponges 8 M256A1 50 trash bags 1 clipboard w/pen 1 stopwatch
Station 4 Rinse	1 sqd l dr 1 PDDA op 2 pump op	2 sprayers	1 M12A1 PDAA 1 3,000 gal tank 3 65-gpm pumps 2 TPU 2 TAP aprons
Station 5 Check	2 Assistance Plt Cdr/CAM op		2 CAM 10 M256A1 20 M8 detector paper 2 AN/PDR 27 2 M8A1 chem alarms
C2	1 Assistance Plt/Cdr 1 Plt Cdr		1 Hummwv/CUCV w/radio 3 NBC marking kits
Total Personnel	20	17	

DECON TEAM EQUIPPED WITH AN M17 LDS.— Decon teams equipped with the M17 LDS will set up the DED differently from M12A PDDA-equipped units. The optimum setup of a DED for an M17 LDS-equipped platoon is shown in figure 6-9.

While this DED configuration is manpower and equipment intensive, it provides for the rapid decontamination of vehicles and equipment (eight vehicles processed per hour) (table 6-9).

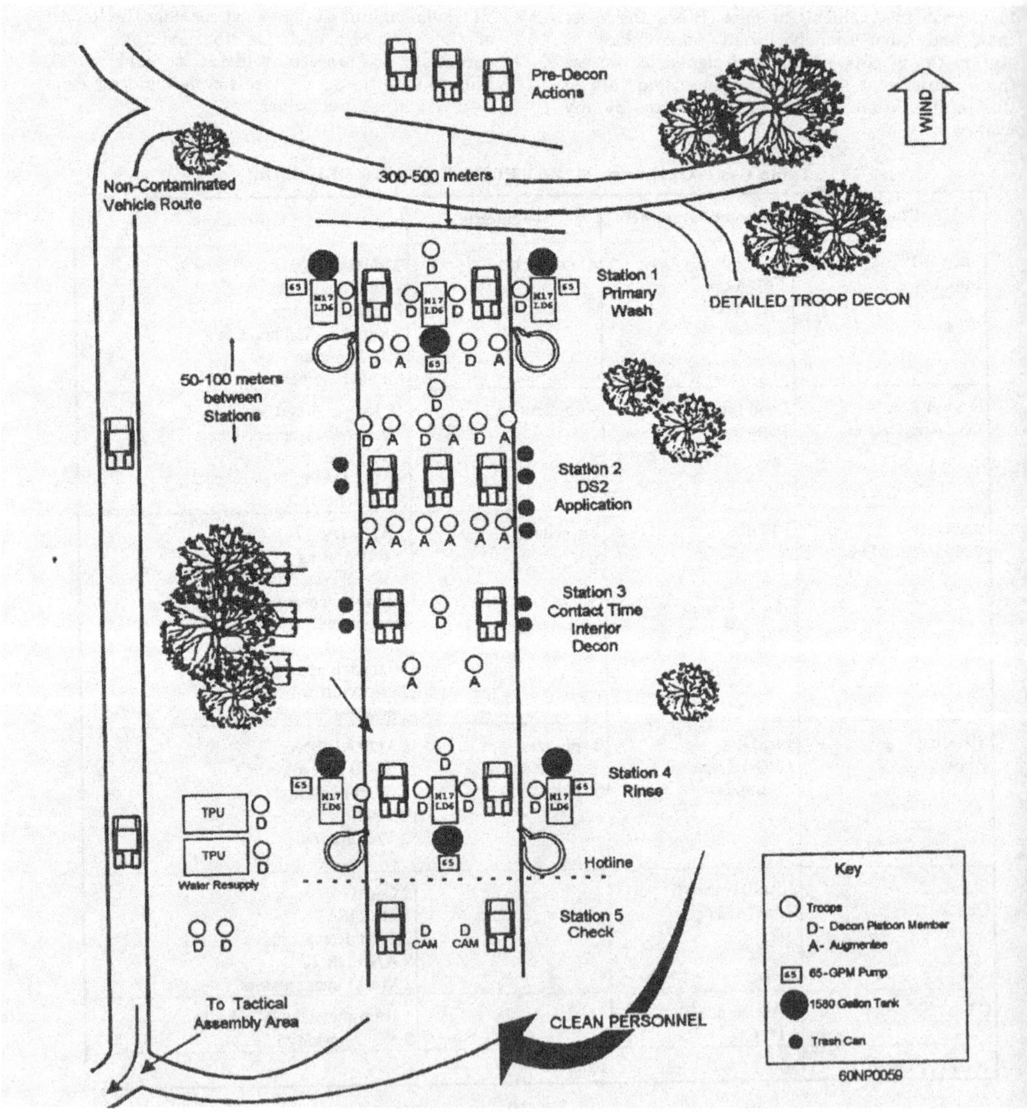

Figure 6-9.—Optimum DED layout for an M17 LDS-equipped unit.

Table 6-9.—Optimum M17 LDS-Equipped DED Setup

Station	Decon Personnel	Augments	Equipment
Station 1 Initial Wash	1 sqd l dr 4 sprayers 2 scrubbers	2 scrubbers	3 M17 LDS 2 1,500-gal tanks 3 65-gpm pumps 6 long-handled brushes 8 TAP aprons Liquid detergent
Station 2 DS2 Application	1 sqd l dr 3 appliers	9 appliers	18 long-handled brushes 9 mops w/extra mop heads 3 30-gal containers 9 M13 DAPs Sufficient DS2
Station 3 Wait/Interior/Decon	1 POIC 2 assistants		2 AN/PDR 27 3 TAP aprons 6 30-gal containers 10 M8 detector paper 30 sponges 8 M256A1 50 trash bags 1 clipboard w/pen 1 stopwatch
Station 4 Rinse	1 sqd l dr 4 sprayers		3 M17 LDS 3 1,500 gal tank 3 65-gpm pumps 4 TAP aprons
Station 5 Check	2 Assistance Plt Cdr/CAM op		2 CAM 10 M256A1 20 M8 detector paper 2 AN/PDR 27 2 M8A1 chem alarms
C2	1 Assistance Plt/Cdr 1 Plt Cdr		1 Hummwv/CUCV w/radio 3 NBC marking kits
Total Personnel	23	11	

This layout uses dual lanes at stations 1,4, and 5 to process two vehicles at once. Since the most time and labor intensive work takes place at station 3, this station is designed to process three vehicles at a time. The processing rate of this configuration will be affected by any work/rest cycle.

DETAILED EQUIPMENT DECON SUMMARY

ALFA company is normally responsible for the setup, the operation, and the closure of the DED portion of the thorough decon operation. The COC will select the DED site. The DED for chemical and biological contamination consists of five stations:

1. Station 1—Initial Wash
2. Station 2—DS2 Application
3. Station 3—Wait/Interior Decon
4. Station 4—Rinse
5. Station 5—Check

DED configurations may vary because of organizational and equipment differences. The optimum configuration provides the maximum output for decon teams at 100 percent personnel and equipment. The setup for a DED equipped with a M12A1 is different from one equipped with a M17. Both the DED and the DTD must be properly closed and marked. Once closed, an NBC 4 report must be sent to higher headquarters.

CLEARING THE THOROUGH DECON SITE

Once all vehicles and personnel from the contaminated unit have been processed through the thorough decon site, the site can be closed. The COC will ensure that all contaminated elements have been processed.

The decon team closes the DED first. Once the DED is closed, the decon team processes through the DTD. After the decon team has processed through the DTD, the DTD is then closed. Once the DTD is closed, the decon team marks the area as a contaminated area and reports its exact location to the COC. The COC informs higher headquarters using an NBC 4 report.

Closing the Detailed Equipment Decon Area

The DED is closed in sequence, starting at station 1. All vehicles, equipment, and nonexpendable supplies are inspected for contamination. If contamination is found, it is decontaminated. The actions at each station are described below.

- **Station 1—Initial Wash.** Spray all vehicles and equipment with hot, soapy water to remove any contamination that could have been transferred during initial wash operations. Drain the water billets or fabric tanks of water. Inspect all equipment and vehicles for contamination, using the appropriate detection equipment. If no contamination is detected, load the equipment on the vehicles. Spread one can of STB into each sump and then cover the sumps. Post NBC hazard markers near the covered sumps.

- **Station 2—DS2 Application.** Throw mops and brushes used in applying DS2 into a sump or bury them. Load unused cans of DS2 and M13 DAPs on a vehicle.

- **Station 3—Wait/Interior Decon.** Inspect any unused supplies and equipment for contamination. If no contamination is detected, load the equipment and supplies on a vehicle. Throw contaminated supplies into the nearest sump.

- **Station 4—Rinse.** Spray all vehicles and equipment with hot, soapy water to remove any contamination that could have been transferred during rinse. Drain the water billets or fabric tanks of water. Inspect all equipment and vehicles for contamination, using the appropriate detection equipment. If no contamination is detected, load the equipment on the vehicles. Spread one can of STB into each sump and then cover the sumps. Post NBC hazard markers near the covered sumps.

NOTE: While DS2 destroys the chemical agents, some by-products created are also toxic.

- **Station 5—Monitor.** Check all the equipment for contamination. If it is not contaminated, load it on a vehicle. If it is contaminated, then decontaminate it. Throw contaminated supplies into the nearest sump.

Move all the vehicles just upwind of station 5 and inspect them again for contamination. If any contamination is detected, the crew will use Ml 1 or M13 DAPs to decontaminate the identified areas. Once the vehicles are decontaminated, all personnel will proceed to the DTD.

Closing the Detailed Troop Decon Area

Once all personnel from the DED have been processed through the DTD, the DTD may be closed. After the last person has exited the DTD, the following steps for closing the DTD are as follows:

- Pickup all used supplies from station 7 and put them in the station 7 sump. Pickup the contamination control line. If tape was used, dispose of it in station 7 sump.

- Move all usable supplies and equipment from all the stations to station 1. Discard unusable supplies from stations 5, 4, and 3 into station 1 sump.

- Decontaminate all supplies and equipment collected at station 1, using the decontaminant and rinse water at station 1. Empty the rinse and decontaminant containers from station 1 into the sump.

- Mark the entire decon area. Remove the overgarments using the MOPP gear exchange technique and dispose of the overgarments in the sump at station 1.

- Move any equipment used to fill the sump upwind of the decon area. Decontaminate the rubber gloves and move all the equipment and supplies in station 1 upwind of the decon area. Keep this equipment and supplies separate from that used to fill the sump.

- Your overboots and gloves may now be contaminated. Remove them. Dig a hole and bury them. Mark the hole and/or area.

APPENDIX I

GLOSSARY OF COMMON MILITARY TERMS

AIR DEFENSE—All defensive measures designed to destroy attacking enemy aircraft or missiles in the earth's envelope of atmosphere or to nullify or reduce the effectiveness of such attack.

AREA OF OPERATIONS (AO)—That portion of an area of war necessary for military operations and for the administration of such operations.

AREA OF RESPONSIBILITY (AOR)—A defined area of land in which responsibility is specifically assigned to the commander of the area for the development and maintenance of installations, the control of movement, and the conduct of tactical operations involving troops under the commander's control, along with parallel authority to exercise these functions.

ARMORED PERSONNEL CARRIER (APC)—A lightly armored, highly mobile, full-tracked vehicle, amphibious and air-droppable, used primarily for transporting personnel and their individual equipment during tiwtical operations. Production modifications or application of special kits permit use as a mortar carrier, a command post, a flame thrower, an antiaircraft artillery chassis, or a limited recovery vehicle.

AVENUE OF APPROACH—An air or ground route of attacking forces of a given size leading to its objective or key terrain in its path.

AVIATION COMBAT ELEMENT (ACE)—One of the four elements of a Marine Air-Ground Task Force (MAGTF). The Aviation Combat Element (ACE) is task-oriented to provide all or a portion of the functions of Marine Corps aviation in varying degrees based on the tactical situation and the MAGTF mission and size. These functions are air reconnaissance, antiair warfare, assault support, offensive air support, electronic warfare, and control of aircraft and missiles. The ACE is organized around an aviation headquarters and varies in size from a composite aircraft squadron to one or more aircraft wing(s). It includes the aviation command (including air control agencies), combat, combat support, and combat service support units required by the situation. Normally, there is only one ACE in a MAGTF.

BARRIER—A coordinated series of obstacles designed or used to channel, direct, restrict, delay, or stop the movement of an opposing force and to impose additional losses in personnel, time, and equipment on the opposing force. Barriers can exist naturally, be man-made, or a combination of both.

BAS—Battalion aid station.

BASE—An area or locality containing installations that provide logistic or other support.

BRIGADE SERVICE SUPPORT GROUP (BSSG)—The BSSG is the task-organized combat service support element of the Marine amphibious brigade (MAB). Personnel and equipment are assigned to the BSSG from the permanent battalions of the force service support group. As required, it may be augmented by combat service support elements from the division or air wing.

CASUALTY STATUS—A term used to classify a casualty for reporting purposes. There are seven casualty statuses: (1) deceased, (2) duty status - whereabouts unknown, (3) missing, (4) very seriously ill or injured, (5) seriously ill or injured, (6) incapacitating illness or injury, and (7) not seriously injured.

COMBAT SERVICE SUPPORT DETACHMENT (CSSD)—A task organization from any combination of combat service support resources. It mayor may not be part of a MAGTF, dependent upon the situation. For example, it could be formed to augment the organic capability of a reinforced infantry battalion conducting an independent operation or to support a squadron located at a remote airfield. The command and control element will normally be provided by the force service support group.

COMBAT SERVICE SUPPORT ELEMENT (CSSE)—One of the four elements of a Marine Air-Ground Task Force (MAGTF). It is task-organized to provide the full range of combat service support necessary to accomplish the MAGTF mission. The CSSE can provide supply, maintenance, transportation, deliberate engineer, health, postal, disbursing, prisoner of war, automated information systems, and exchange. The CSSE varies in size from a MAU service support group (MSSG) to a force service support group (FSSG). Normally, there is only one Combat Service Support Element in a MAGTF.

COMBAT SERVICE SUPPORT OPERATION CENTER (CSSOC)—The agency that controls and coordinates the day-to-day operations of the combat support elements.

COMBAT SUPPORT ELEMENTS (CSE)—Those elements whose primary missions are to provide combat support to the combat forces and which are a part, or prepared to become a part, of a theater, command, or task force formed for combat operations.

COMMAND AND CONTROL—The exercise of authority and direction by a properly designated commander over assigned forces in the accomplishment of the mission. Command and control functions are performed through an arrangement of personnel, equipment, communications, facilities, and procedures used by a commander in planning, directing, coordinating, and controlling forces and operations in the accomplishment of the mission.

COMMAND ELEMENT (CSE)—One of the four elements of a Marine Air-Ground Task Force (MAGTF). It is the MAGTF headquarters. The Command Element is a permanent organization composed of the commander, the general or executive and special staff sections, the headquarters section, and requisite communications and service support facilities. The command element provides command, control, and coordination essential for effective planning and execution of operations by the Aviation Combat Element (ACE), the Ground Combat Element (GCE), and the Combat Service Support Element (CSSE). There is only one command element in a MAGTF.

CONTAMINATION—The deposit and/or absorption of radioactive material or biological or chemical agents on and by structures, areas, personnel, or objects.

CONVOY ESCORT—An escort to protect a convoy of vehicles from being scattered, destroyed, or captured.

DEAD SPACE—An area within the maximum range of a weapon, a radar, or an observer that cannot be covered by fire or observation from a particular position because of intervening obstacles, the nature of the ground, the characteristics of the trajectory, or the limitations of the pointing capabilities of the weapons.

DECONTAMINATION—The removal or neutralization of hazardous levels of chemical, biological or radiological contamination from personnel and material.

DEFENSE AREA—For any particular command, the area extending from the forward edge of the battle area to its rear boundary. It is here that the decisive defensive battle is fought.

DEFENSE IN DEPTH—The siting of mutually supporting defense positions designed to absorb and progressively weaken attack, to prevent initial observations of the whole position by the enemy, and to allow the commander to maneuver his reserve.

DETAILED EQUIPMENT DECON—Process of removing or neutralizing contamination on interior and exterior surfaces of unit equipment to negligible risk levels to allow MOPP level reduction for extended periods.

DETAILED TROOP DECON—Process of decontaminating individual fighting equipment to negligible risk levels; removing contaminated MOPP gear including protective masks; decontaminating protective masks; and monitoring personnel equipment for decon effectiveness. This is done to reduce MOPP levels for extended periods.

FINAL PROTECTIVE FIRE (FPF)—An immediately available prearranged barrier of fire designed to impede enemy movement across defensive lines or areas.

FIRE PLAN—A tactical plan for using the weapons of a unit or formation so that their fire will be coordinated.

FIREPOWER—The amount of fire which may be delivered by a position, a unit, or a weapon system.

FORCE SERVICE SUPPORT GROUP (FSSG)—The FSSG is a permanently structured organization whose mission is to provide CSS for the Marine amphibious force (MAF). It is designed to support one division and one air wing. If supporting a force of greater size, additional assets are necessary to augment its capabilities.

GROUND COMBAT ELEMENT (GCE)—One of the four elements of a Marine Air-Ground Task Force (MAGTF). It is task-organized to conduct ground operations. The GCE is constructed around an infantry unit and varies in size from a reinforced infantry battalion to one or more reinforced Marine division(s). The GCE also includes appropriate combat support and combat service support units. Normally, there is only one Ground Combat Element in a MAGTF. Although permanently structured with eight functional battalions, task organizations from those battalions would normally support MAF operations over a wide geographic area.

HASTY DECON OPERATION—A decon operation that consists of two techniques, the MOPP gear exchange and the vehicle wash down.

HELICOPTER LANDING SITE—A designated subdivision of a helicopter landing zone in which a single flight or wave of assault helicopters land to embark or disembark troops and/or cargo.

HELICOPTER LANDING ZONE (HLZ)—A specified ground area for landing assault helicopters to embark or disembark troops and/or cargo. A landing zone may contain one or more landing sites.

HOT LINE—A real or imaginary line that separates contaminated from uncontaminated areas.

INTELLIGENCE—(1) The product resulting from the collection, processing, integration, analysis, evaluation, and interpretation of available information concerning foreign countries or areas and (2) the information and knowledge about an adversary obtained through observation, investigation, analysis, or understanding.

KEY TERRAIN—Any locality, or area the seizure or retention of which affords a marked advantage to either combatant.

KILLING ZONE—An area in which a commander plans to force the enemy to concentrate so as to destroy him with conventional weapons or the tactical employment of nuclear weapons.

LANDING ZONE (LZ)—Any specified zone used for the landing of aircraft.

MAIN BATTLE AREA—That portion of the battlefield in which the decisive battle is fought to defeat the enemy. For any particular command, the main battle area extends rearward from the forward edge of the battle area to the rear boundary of the command's subordinate units.

MARINE AIR-GROUND TASK FORCE (MAGTF)—A Marine Air-Ground Task Force is a task organization of Marine forces (division, aircraft wing, and service support groups) under a single command and structured to accomplish a specific mission. The Marine Air-Ground Task Force components will normally include command, ground combat, aviation combat, and combat service support elements (including Navy Support Elements).

MARINE AIRCRAFT GROUP (MAG)—The MAG is usually administratively and tactically structured by aircraft category as being either a helicopter group or a fixed-wing group. Composite MAGs may also be formed for specific missions or unique organizational/geographic considerations. Each MAG has a headquarters and maintenance squadron (H&MS). With a source of supply, the MAG is the smallest aviation unit capable of self-sustaining independent operations.

MARINE AIRCRAFT WING (MAW)—The MAW is the highest level aviation command in the Fleet Marine Force (FMF). Each wing is capable of supporting one Marine division. The MAW is task-organized to provide a flexible and balanced air combat organization capable of providing the full range of combat air operations in a variety of areas without the requirement of prepositioned support, control, and logistics facilities. Only the wing has the inherent capability of performing all six aviation functions.

MARINE EXPEDITIONARY BRIGADE (MEB)—A task organization which is normally built around a regimental landing team, a previsional Marine aircraft group, and a logistics support group. It is capable of conducting amphibious assault operations of a limited scope. During potential crisis situations, a Marine Expeditionary Brigade may be forward deployed afloat for an extended period in order to provide an immediate combat response.

MARINE EXPEDITIONARY FORCE (MEF)—The Marine Expeditionary Force, the largest of the Marine air/ground task forces, is normally built around a division/wing team, but can include several divisions and aircraft wings, together with an appropriate combat service support organization. The Marine expeditionary force is capable of conducting a wide range of amphibious assault operations and sustained operations ashore. It can be tailored for a wide variety of combat missions in any geographic environment.

MARINE EXPEDITIONARY UNIT (MEU)—A task organization which is normally built around a battalion landing team, reinforced helicopter squadron, and logistic support unit. It fulfills routine forward afloat deployment requirements, provides an immediate reaction capability for crisis situations, and is capable of relatively limited combat operations.

MARINE EXPEDITIONARY UNIT (SPECIAL OPERATIONS CAPABLE)—A forward-deployed, embarked U.S. Marine Corps unit with enhanced capability to conduct special operations. The Marine expeditionary unit (special operations capable) is oriented toward amphibious raids, at night, under limited visibility, while employing emission control procedures. The Marine expeditionary unit (special operations capable) is not a Secretary of Defense-designated special operations force but, when directed by the National Command Authorities and/or the theater commander, may conduct hostage recovery or other special operations under extreme circumstances when designated special operations forces are not available. Also called MEU (SOC).

MARINE WING SUPPORT GROUP (MWSG)—An administrative command that provides all essential aviation ground support (AGS) requirements for elements of the MAW. The MWSG is capable of supporting two fixed-wing and two rotary-wing airfields per MAW.

MAU SERVICE SUPPORT GROUP (MSSG)—A task-organized combat service support element of the Marine amphibious unit (MAU). Like the BSSG, it draws personnel and equipment from the permanent battalions of the force service support group. As required, it may be augmented by combat service support elements from the division or aircraft wing.

MISSION ORIENTED PROTECTIVE POSTURE—A flexible system for protection against NBC contamination. This posture requires personnel to wear only that protective clothing and equipment (MOPP gear) appropriate to the threat, work rate imposed by the mission, temperature, and humidity. There are five levels of MOPP (zero through 4). MOPP 4 offers the most protection but also degrades mission performance the most.

MOPP GEAR—Combat clothing and equipment used to operate in an NBC environment.

MOPP GEAR EXCHANGE—Process of decontaminating individual fighting equipment, including the protective mask and hood, removing the contaminated MOPP gear, and putting on new MOPP gear.

NEGLIGIBLE RISK LEVELS—Levels of contamination that will cause mild incapacitation among no more than 5 percent (for chemical and biological contamination; 2.5 percent nuisance effect for radiological contamination) of the unprotected troops who operate for 12 continuous hours within 1 meter of a contaminated surface.

OPERATIONAL CONTROL (OPCON)—The authority delegated to a commander to perform those functions of command over subordinate forces involving the composition of subordinate forces, the assignment of tasks, the designation of objectives, and the authoritative direction necessary to accomplish the mission. Operational control includes directive authority for joint training. Operational control should be exercised through the commanders of assigned normal organizational units or through the commanders of subordinate forces established by the commander exercising operational control. Operational control normally provides full authority to organize forces as the operational commander deems necessary to accomplish assigned missions and to retain or delegate operational control or tactical control as necessary. Operational control may be limited by function, time, or location. It does not, of itself, include such matters as administration, discipline, internal organization, and unit training.

OPERATOR'S SPRAY DOWN—Process of applying decontaminant onto unit equipment control surfaces to stop contamination from soaking into surfaces.

PARTIAL DECONTAMINATION—The removal or neutralization of all visible or detectable contamination from individual clothing and equipment and from those surfaces of equipment that operators or crew members must contact to perform their mission-vehicle entry and exit routes.

PASSIVE AIR DEFENSE—All measures, other than active air defense, taken to minimize the effectiveness of hostile air action. These measures include deception, dispersion, and use of protective construction.

PASSIVE DEFENSE—Measures taken to reduce the probability of and to minimize the effects of damage caused by hostile action without the intention of taking the initiative.

PERSONAL WIPE DOWN—Process of removing or neutralizing contamination from the individual's equipment including the protective mask, hood, gloves, rifle, and helmet to stop contamination spread and to stop contamination from penetrating into equipment surfaces.

PHYSICAL SECURITY—That part of security concerned with physical measures designed to safeguard personnel; to prevent unauthorized access to equipment, installations, material, and documents; and to safeguard them against espionage, sabotage, damage, and theft.

POWER-DRIVEN DECON EQUIPMENT—Any of several different kinds of pump and heater units capable of spraying heated water or steam. Soap and decontaminants can also be mixed and sprayed through these units in most cases, such as the M12 power-driven decon apparatus and the XM17 SANATOR lightweight decon system.

PROVISIONAL MOBILE SECURITY PLATOON (PMSP)—Unit of the provisional security forces which provides the rear area security coordinator with a quick reaction capability in support of the RAS mission.

PROVISIONAL SECURITY FORCES (PSF)—Units available to the rear area security coordinator for supplementing local defense efforts and to give assistance to the military police in the performance of RAS missions.

PYROTECHNIC—A mixture of chemicals which when ignited is capable of reacting exothermically to produce light, heat, smoke, sound or gas, and may also be used to introduce a delay.

REAR AREA OPERATIONS CENTER (RAOC)—The agency responsible for planning, coordinating, directing, and monitoring rear area security.

REAR AREA SECURITY (RAS)—The measures taken before, during, and/or after an enemy airborne attack, sabotage action, infiltration, guerrilla action, and/or initiation of psychological or propaganda warfare to minimize the effects thereof.

REAR AREA SECURITY COORDINATOR (RASC)—That person responsible for planning, coordinating, and directing the RAS effort. The RASC is usually the CSSE commander but can be the ACE commander.

RECONSTITUTION—The rest, refitting, maintenance, and replacement necessary to restore a military unit to its full capability, often after it has been depleted by military operations.

RATE OF FIRE—The number of rounds fired per weapon per minute.

RATE OF MARCH—The average number of miles or kilometers to be travelled in a given period of time, including all ordered halts. It is expressed in miles or kilometers in the hour.

SKIN DECON TECHNIQUE—Process of removing or neutralizing contamination on the skin within 1 minute of contamination to prevent it from penetrating into skin.

SITUATION MAP—A map showing the tactical or the administrative situation at a particular time.

SITUATION REPORT (SITREP)—A report giving the situation in the area of a reporting unit or formation.

SMALL ARMS—Man portable, individual, and crew-served weapon systems used mainly against personnel and lightly armored or unarmored equipment.

SPOT REPORT—A concise narrative report of essential information covering events or conditions that may have an immediate and significant effect on current planning and operations that is afforded the most expeditious means of transmission consistent with requisite security.

STANDING OPERATING PROCEDURE (SOP)—A set of instructions covering those features of operations which lend themselves to a definite or standardized procedure without loss of effectiveness. The procedure is applicable unless ordered otherwise.

TABLE OF ALLOWANCE (TOA)—An equipment allowance document which prescribes basic allowances of organizational equipment and provides the control to develop, revise, or change equipment authorization inventory data.

TACTICS—(1) The employment of units in combat and (2) the ordered arrangement and maneuver of units in relation to each other and/or to the enemy in order to use their full potentialities.

TAP APRON—Toxicological agent protective apron.

TARGET OF OPPORTUNITY—A target visible to a surface of air sensor or observer which is within range of available weapons and against which fire has not been scheduled or requested.

TERRAIN ANALYSIS—The collection, analysis, evaluation, and interpretation of geographic information on the natural and man-made features of the terrain, combined with other relevant factors, to predict the effect of the terrain on military operations.

VEHICLE WASH DOWN—Process of flushing contamination off equipment surfaces to limit spread, reduce overall amounts of contamination, and speed weathering.

APPENDIX II

OVERLAY TECHNIQUES

1. GENERAL

Overlays provide a rapid and easily understood means by which the commander or his staff may express an operational plan, concept, or friendly or enemy situation. Standardization of technique is essential if tactical information is to be relayed without misunderstanding. Guidelines for the pictorial representation of tactical situations are established in this appendix.

2. MILITARY SYMBOLS

a. Colors.—Colors in conjunction with military symbols denote the following:

(1) Blue or Black.—Friendly units and activities.

(2) Red. —Enemy units and activities. If this color is not available, enemy symbols are outlined with double black lines.

(3) Yellow.—Friendly or enemy areas of chemical, biological, or radiological contamination.

(4) Green.—Man-made obstacles.

b. Units and Installations

(1) Geometric figures form the basic symbols to represent units and installations. Future or proposed locations of units or installations are shown by broken lines. Examples of the more common figures areas follows:

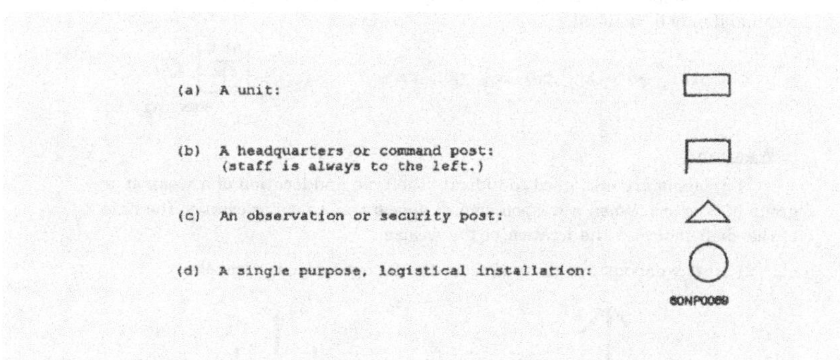

(2) To show the size of a unit, the appropriate size indication is placed on top of the basic symbols as follows:

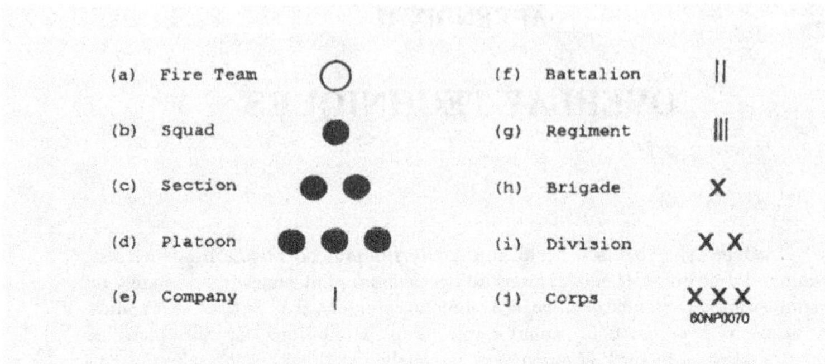

(3) To showthe type unit being represented, a symbol is placed inside the basic figure as follows:

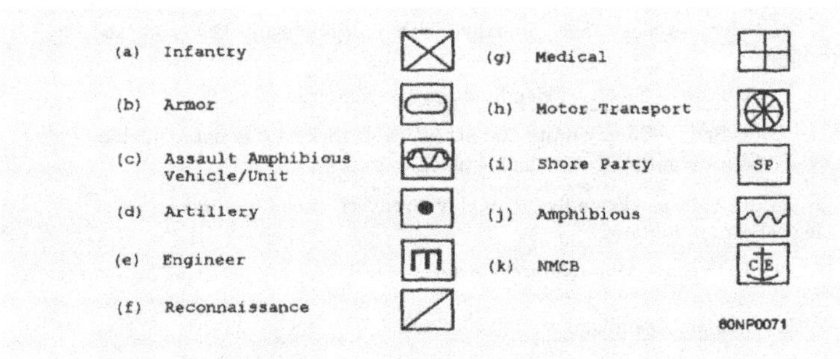

(4) To indicate a particular unit or installation, place the unit's own designation (in accordance with its size symbol) to the left of the symbol with higher echelons of command to the right of the symbol. Slashes separate command echelons. To avoid cluttering the symbol, known units in the chain of command may be omitted.

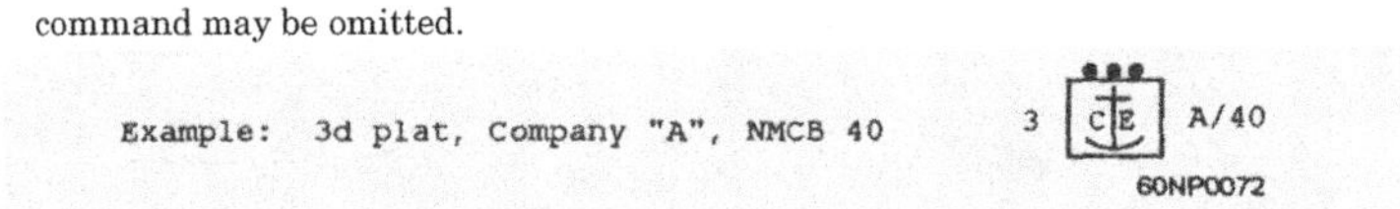

c. <u>Weapons</u>

(1) Symbols are also used to indicate the type and location of a weapon or group of weapons. When a weapon symbol appears on a map or overlay, the base of the shaft indicates the location of the weapon.

(2) Most weapons are derived from the following basic symbols:

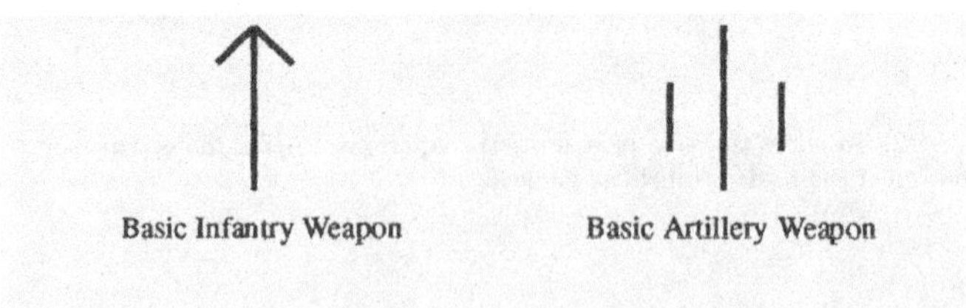

(a) If the weapon has a high trajectory, a ○ is placed at the base of the weapon.

(b) A weapon which is a flat trajectory, antitank weapon has a ∧ placed at the base of the shaft.

(c) If the weapon is primarily for air defense, a ⌓ is placed at the base of the shaft.

(d) A weapon which is a rocket projector or launcher has a ⩓ placed at the head of the shaft.

(e) If the weapon is also a tracked, self-propelled vehicle, a ◇ is placed below the weapon symbol.

(3) Generally, the number and caliber of weapons are indicated by placing the number of weapons to the left of the symbol and the caliber to the right of the symbol.

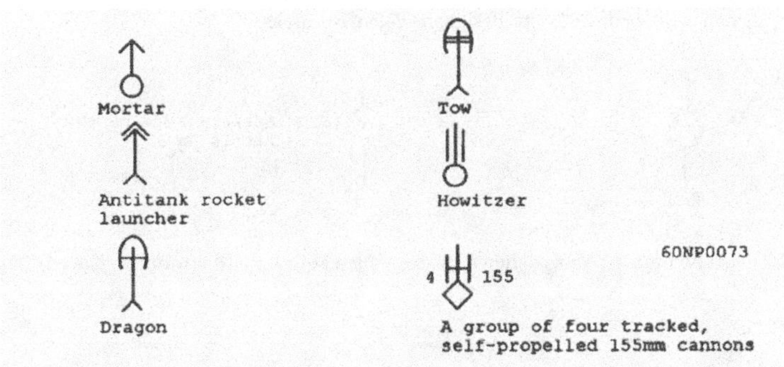

3. CONTROL MEASURES

a. Lines of control or coordination are drawn and labeled as shown below:

(1) Fire support coordination line	FSCL ____________ FSCL
(2) Restrictive fire line	RFL ____________ RFL
(3) Light line	LL ____________ LL
(4) Line of departure	LOD ____________ LOD
(5) Line of departure is present positions	LOD/PP ⌒⌒⌒⌒ LOD/PP
(6) Line of departure is forward friendly dispositions	LOD/FFD ⌒⌒⌒⌒ LOD/FFD
(7) Phase line with code name	PL ____________ PL GREEN GREEN

(8) Coordinated fire line CFL__________CFL

(9) Probable line of deployment PLD . . . . PLD

(10) Final coordination line FINAL CL__________FINAL CL

b. Control points are drawn on the selected terrain feature and identified as follows:

(1) A checkpoint consists of a circle enclosing a selected terrain feature with a number, letter, or code name placed inside the circle.

(2) A coordinating point is shown by drawing a circle on the selected terrain feature and placing an "X" in the center. Coordinating points are used in conjunction with boundaries to designate defensive areas.

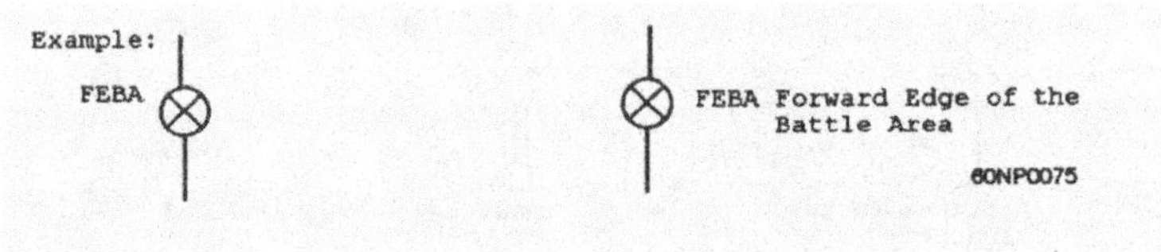

(3) A contact point is shown by drawing a square with a number placed inside.

4. FIRE PLANS

a. Sector of Fire

(1) Representation of a sector of fire is shown by two arrows composed of broken lines:

(2) A weapon symbol is normally used in conjunction with the symbol for a sector of fire. The base of the symbol indicates the weapon's position.

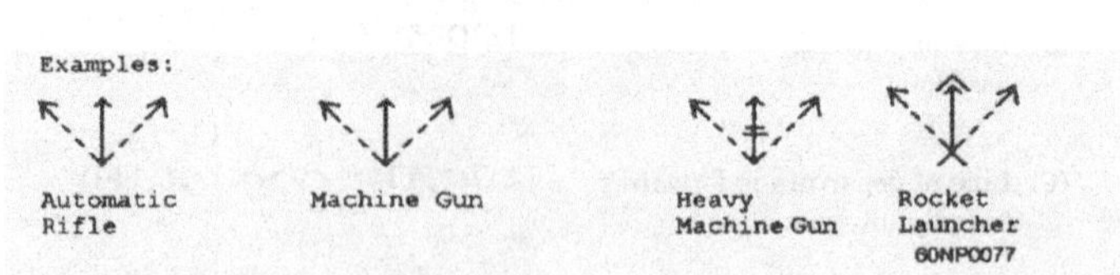

b. <u>Direction of Fire</u>

(1) A principal direction of fire is represented by a solid arrow. To prevent confusion with similar symbols, the symbol representing a principal direction of fire is always shown together with the appropriate weapon symbol.

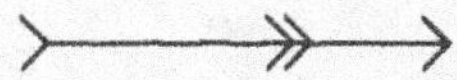

Principal direction of fire for an antitank rocket launcher

(2) Symbols for sectors of fire and principal directions of fire are often combined.

Example:

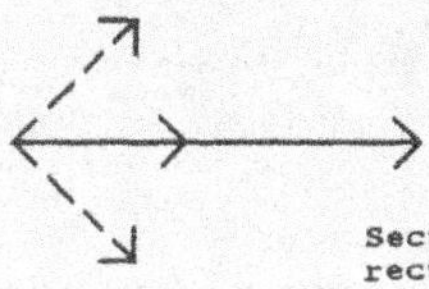

60NP0079 Sector of fire and principal direction of fire for a machine gun.

(3) A special principal direction of fire symbol is used to indicate final protective lines. Heavily shaded portions along the principal direction of fire symbol indicate areas of grazing fire.

Example:

60NP0080 Final protective line for a single machine gun.

T

he final protective line symbol is usually combined with the sector of fire and weapon symbols.

Examples:

A final protective line and sector of fire for a single machine gun (machine gun team).

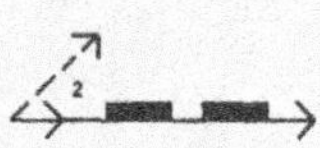

A final protective line and sector of fire for a machine gun squad (two machine guns). 60NP0081

c. <u>Targets</u>

(1) Point targets of less than 100 yards

(2) Linear targets

60NP0083

(3) Area targets

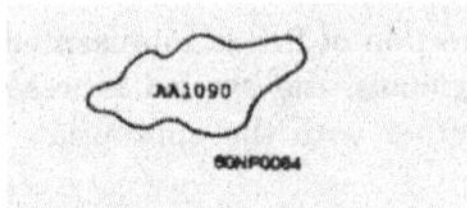

(4) Rectangular targets

d. <u>Final Protective Fire</u>

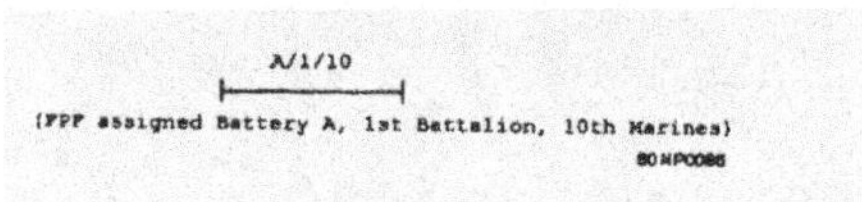

APPENDIX III

CHARACTERISTICS OF TOA WEAPONS FOR AN NMCB

WEAPON	AIR DET TOA	AIR ECH TOA	MAX EFFECTIVE RANGE/ METERS	RATE OF FIRE RDS PER MINUTE
Mortar 60MM M224	2	4	3,490	Sustained 15 Rapid 30
M16A2E3	76	574	800	Semi 45 Auto 90
Grenade Launcher M203	6	42	Area Target 350 Point Target 150	X
Pistol .45 Cal M1911A1	20	155	45	Sustained 10 Rapid 21-28
Shotgun M870	3	21	45	X
Machine Gun 50 Cal M2	2	4	1,830	Sustained 40 or less Rapid more than 40
Machine Gun 40MM MK19	2	4	1,500	Sustained 40 Rapid 60 or more
Machine Gun 7.62MM M60E3	4	12	Tripod 1,110 Bipod 750	Sustained 100 Rapid 200
Anti-Tank Weapon 84MM M136 HEAT	X	X	300	X

APPENDIX IV

DECONTAMINANTS

Decontaminants in this appendix are grouped in tables according to their classification of standard, nonstandard, and natural. See Tables A-1 through A-3. Each decontaminant is identified by name and NSN (where applicable). Some include pictures of the decontaminant or its container. The use of each decontaminant is listed as nuclear, biological, and/or chemical, and brief directions are given on how to prepare each decontaminant. The tables also list any appropriate cautions for each decontaminant. A remarks column gives information on application, coverage, subsitute solutions, and the related effectiveness of the decontaminants.

Table A-1.—Standard decontaminants
(Decontaminants most often used and available in supply system)

Decontaminant	Use	Remarks	Cautions	Preparation
Decontaminating Solution No. 2 (DS2)  60NP0047 DS2 in a 1-1/3 quart can NSN 6850-00-753-4827  09NP0048 DS2 in 14-liter container for M13 DAP NSN 4230-01-136-8888  80NP0049 DS2 in a 5 gallon pail NSN 6850-00-753-4870	Bio Cml	Effective against all known toxic chemical agents and biological materials (except bacterial spores) if sufficient contact time is allowed. Allow to remain in contact with contaminated surface for approximately 30 minutes. Rinse off with water. Recheck for contamination. Can be used at temperatures from - 15°F Used with the ABC-M11 1-1/2 quart portable decon apparatus, M13 DAP, or can be applied with brooms and swabs. Most effective when application is accompanied by scrubbing action.	Extremely irritating to the eyes and skin. Protective mask and rubber gloves must be worn. If DS2 contacts skin, wash the area with water. Do not inhale vapors. Will cause a green to black color change upon contact with ABC-M8 detector paper and cause a false/positive with M9 paper. Ignites spontaneously on contact with STB and HTH. Avoid spilling DS2 on chemical protective overgarment. Combustible. Do not confuse with fire extinguisher. DS2 is a combustible liquid with a flash point of 160°F. Spraying DS2 onto heated surfaces above 168°F will ignite the DS2. Do not use on M17-series mask (damages mylar diaphragm in voicemitter assembly). Corrodes aluminum, cadmium, tin, and zinc; softens leather. May soften, remove, or discolor paint. Rinse well after use and oil metal surfaces Ineffective against bacterial spores.	No mixing is required. Issued in ready-to-use solutions.

Table A-1.—Standard decontaminants—Continued
(Decontaminants most often used are available in supply system)

Decontaminant	Use	Remarks	Cautions	Preparation
Supertropical Bleach (STB)  80NP0050 NSN 6850-00-297-6653	Bio Cml	Effective against Lewisite, V and G agents, and biological agents. Allow to remain in contact with contaminated surface for at least 30 minutes, then wash off with clear water.	Ignites spontaneously on contact with liquid blister agent or DS2. Gives off toxic vapors on contact with G agent, Not recommended for ship use. Top deck storage only. Corrosive to most metals and injurious to most fabrics (rinse thoroughtly and oil metal surfaces). STB mixtures (dry and slurry) do not effectively decon mustard if it has solidified at low temperatures. Porous surfaces may require several applications. Should not be inhaled or come in contact with the skin. Protective mask or other respiratory protection device should be worn when preparing slurry. Store in unheated warehouse isolated from combustibles and metals subject to corrosion.	**Slurry Paste**—Mix one 50-lb drum of STB with 6 gal of water. Slurry paste consists of approximately equal parts (by weight) of STB and water. **Dry Mix**—2 shovels STB to 3 shovels earth or inert material (such as ashes). **Slurry Mix**—Chemical-Slurry mix will consists of 40 parts STB to 60 parts water (by weight). To mix in M12A1 use 1,300 lb. STB, 225 gal water, 12-1/2 lb antiset, 24 oz antifoam. **Biological**—Slurry mix will consist of 7 parts STB to 93 parts water (by weight). To mix in M12A1 use 150 lb STB, 225 gal water, 1-1/2 lb antiset, 24 oz antifoam. **Camouflage**—Lampblack or dye mixes maybe added for camouflage. No mixing.
Mask Sanitizing Solution	Bio Cml	Used on previously cleaned masks with filter elements removed. Place mask face up. Attach canteen to mask at the drinking tube. Drain one canteen full of sanitizing solution through the mask. Follow with two canteens of clean water as a rinse. Immerse mask and outserts in sanitizing solution. Agitate for 5 minutes. Rinse twice in clear water, agitating 2 or 3 minutes each time. Dry all parts and reassemble mask.	1 gallon of solution needed for every ten masks.	Fill standard plastic canteen to shoulder with water. Add one 0.5 gram tube calcium hypochlorite from water purification kit (NSN 6810-00-266-6979). Cover canteen and shake vigorously for 30 seconds. Mix bulk quantities as follows. Add 2.0 grams (.08 oz) of calcium hypochlorite from 6-oz jar (NSN 6810-00-255-0471) to 1 gallon of water.
Soap and Detergents Detergent, general purpose, liquid (NSN 7930-00-282-9699)	Nuc Bio Cml	Scrub or wipe contaminated surfaces with hot, soapy water solution or immerse item in the solution.	Soaps and detergents are effective in physically removing contamination. However, casualty-producing levels of contamination may remain in the runoff water which must be considered contaminated.	Mix 75 lb of powdered soap in 350 gal of water. If powdered soap is not available, bar laundry soap may be used (75 lb of soap, cut into 1-inch pieces and disolved in 350 gal hot water). For smaller amounts of soap solution, use a ratio of approximately 1 lb soap per gal of water. Mix 2-pints detergent to 450 gallons water in M12A1 PDDE.

Table A-2.—NonStandard Decontaminants
(Decontaminants not frequently used)

Decontaminant	Use	Remarks	Cautions	Preparation
Oxidizing Agents (potassium permanganate, potassium or sodium dichromate, nitric acid, or aqua regia)	Nuc	Effective in dissolving surfaces containing adsorbed or absorbed radioactive contamination. Dip into or coat surface with oxidizing agent. Exposure must be limited due to corrosive nature of solution. Rinse thoroughly with water and detergent and then with clear water.	Extremely corrosive. Use only under the supervision of an individual trained in their use. Neoprene or rubber protective apron, gloves, boots, and safety glasses must be worn. (Rubber offers only limited protection.)	Aqua regia is prepared by mixing 3 parts of concentrated hydrocloric acid and 1 part concentrated nitric acid. Other oxidizing agents do not require mixing.
Complexing (Chelating) Agents (versene, sequesterene, citric acid, sodium citrate, tartaric acid, sodium tartrate, oxalic acid, sodium oxalate, orthophosphoric acid, and similar agents)	Nuc	Aids in removal of contamination that is absorbed on surfaces. Apply as a film over contaminated surface using PDDA, fire fighting apparatus, or tree or garden sprayer. Allow 30 minutes contact time and flush with water.	Aids in physical removal of contamination but does not neutralize contamination. Runoff residue will be contaminated.	Mix 3 to 5 percent of agent (by weight) in water.
Iodine Water Purification Tablets	Bio	Where boiling of drinking water is impractical, two iodine tablets per canteen of water (proper instructions on container) are effective against most biologival agents.	None	None
Disinfectant, Chlorine, Food Serve (NSN 6840-00-270-8172)	Bio	Effective for decon of utensils, mess gear, the exteriors of sealed containers, and food products that can withstand soaking. Dispose of any food or vegetable that is damaged and any outer leaves that are bruised or torn. Do not cut or peel fruits and vegetable before disinfecting them. Leave items in solution for 30 minutes and stir occasionally to insure that surfaces are kept thoroughly wet. Utensils may be disinfected by immersing in solution for 30 seconds. Rinse thoroughly in potable water. Do not use solutions more than once. If this disinfectant is not available, an emergency solution prepared by mixing at least one level mess kit spoonful of calcium hypochlorite (water disinfecting powder) to each 10 gal of water. If liquid chlorine bleach is available, it may be used. About one-third canteen cup of 5 percent chlorine bleach to each 10 gal of water will produce the same disinfecting strength. Fresh solutions must be made for rinsing and disinfecting utensils for each 100 persons.	None	Dissolve one package of disinfectant in 20 gal of warm potable water (100°F)

Table A-2.—NonStandard Decontaminants—Continued

Decontaminant	Use	Remarks	Cautions	Preparation
Formalin (formaldehyde)	Bio	Effective against all micro-organisms, to include bacterial spores. Recommended as interior decontaminant for relatively closed areas. Allow vapors to remain 16 hours in a closed structure, then aerate until odor is no longer objectionable. Optimum conditions for spraying are 70° to 80°F. 85 percent relative humidity. The minimum effective relative humidity is 70 percent. The minimum effective temperature is 60°F. at which the exposure time should be increased to 24 hours. Apply as vapor from standard insecticide sprayers or vaporize by heat or bubbling steam through pan of decontaminant.	Formalin vapors are very toxic. Will curl and discolor paper. Will leave white residue. Up to 72 hours aeration may be required. A self-contained breathing apparatus should be worn when remaining for more than a few minutes in a building containing formalin vapors. Personnel handling or spraying formalin should wear impermeable protective clothing. Personnel entering an area containing formalin vapor for only a few minutes should— • Wear protective mask. • Wear washable outer clothing, fastened to prevent vapor from entering at wrist, ankles, or neck. • Remove outer clothing after emerging from vapor. • Shower and put on clean clothing as soon as possible. Vapors of formalin are not flammable; open flame should not be used for vaporizing when methanol has been added to formalin. When steam is used, source of steam should be outside area being decontaminated.	No mixing required. However, less residue remains and less aeration is required if mixture of 5 parts formalin and 3 parts methanol are used. (Use this mixture at rate of 4/5 qt per 1,000 cu ft of space.)
Detrochlorite	Bio	A thickened bleach useful on vertical surfaces. Apply by means of a PDDA. Allow 30 minutes contact time, then rinse with water. Coverage is 1 gal per 8 sq yd.	Very corrosive. See "Preparation" column.	Mix by weight 19.3 percent diatomaceous earth, 0.5 percent anionic wetting agent, 2.9 percent calcium hypochlorite (70 percent available chlorine), 77.3 percent water. Mix wetting agent and diatomaceous earth with water before adding the calcium hypochlorite. Mixing the wetting agent and calcium hypochlorite in a dry undiluted state may cause an explosion.
Paracetic Acid (PAA)	Bio	Effective against all micro-organisms to include bacterial spores. Allow 10 minutes contact time. Wipe item with rag or swab, remove excess acid, and aerate 10 to 15 minutes, or until no objectionable odor remains. Immerse small items for 10 minutes, remove excess acid, and aerate until no objectionable odor remains.	Fumes highly irritating. Prolonged exposure will damage most materials. 40 percent solution has low flash point (105°F); 3 percent solution is nonflammable. Will cause burns and blister skin Must be sorted in original containers under refrigeration to prevent decomposition. Protective clothing and mask are required.	Available as 40 percent solution. Mix 1 qt to 3-1/2 gal of water (add peracetic acid to water.)

Table A-2.—NonStandard Decontaminants—Continued

Decontaminant	Use	Remarks	Cautions	Preparation
Paracetic Acid (PAA)—Continued	Bio	SAME	A violent explosion may result if heavy metal ions come in contact with peracetic acid. Prolonged exposure will corrode iron and deteriorate rubber, Plastics, and leather.	SAME
Ethylene Oxide (ETO)	Bio	Effective against all micro-organisms to include spores. Apply 30 lb for every 1,000 cu ft. Allow 6 hr contact time (contact time must be doubled for each 20°F drop in temperature below 75°F). Airtight enclosure required.	ETO is flammable. Not recommended for interior use (see carboxide below).	None
Carboxide	Bio	Carboxide is a mixture of ethylene oxide and carbon dioxide that is effective against all micro-organisms to include spores. Apply 30 lb for every 1,000 cu ft. Allow 12 hr contact time (contact time must be doubled for each 20°F drop in temperature below 75°F). Carboxide is nonflammable and is recommended for interior use. Airtight enclosure required.	Will blister skin. Items worn next to skin must be aerated 18 to 24 hr.	None
Hyamine (Benzethonium chloride)	Bio	Effective against all bacteria. Allow 5 to 30 minutes contact time.	Very toxic. Estimated fatal dose to man 1 to 3 grams. Care should be taken when mixing to avoid inhalation of powder. Not to be used on ships.	Use a 0.1 to 1 percent solution (1 lb hyamine for every 12 gal of water yields 1 percent solution).
Sodium Hypochlorite Solution (household bleach)	Bio Cml	Effective against blister and V agents and all biological materials. Reacts rapidly (within 5 minutes) with blister and V agents. Allow 10 to 15 minutes contact time for biological materials. Possible sources—commercial laundry (19 to 14 percent solution) or food store (5 percent solution such as Purex or Clorox® Apply undiluted with brooms, brushes, or swabsa. Preferred decontaminant for ship use. For ship use, a 5 to 1 concentration is recommended. Limited storage problem.	Undiluted, it is harmful to skin and clothing. Remove from skin and clothing by flushing with water. Corrosive to metals unless rinsed, dried, and lubricated after decon. Store in cool place.	No mixing required for chemical decon. For biological decon, dilute by adding 2 parts bleach to 10 parts water. For decon of cotton clothing and untensils, bleach should be diluted 1/2 cup bleach to 1 gal water. Dilute half and half with water and spray from PDDE.

Table A-2.—NonStandard Decontaminants—Continued

Decontaminant	Use	Remarks	Cautions	Preparation
Calcium Hypochlorite (HTH) (high test bleach, HTB, or high test hypochlorite) NSN 6810-264-6591 (25 lb) NSN 6810-664-0402 (100 lb)	Bio Cml	Effective against Lewisite, V agents, and all biological materials including bacterial spores. Reacts rapidly (within 5 minutes) with mustard and Lewisite. Allow 15 minutes contact time for biological materials. Faster acting than STB. Can be used as a dry mix or a slurry. Not recommended for ships. Top deck storage only. Not allowed in slurry or dry form in holds of vessels. Possible sources, commercial laundry, drug store, or chemical firm.	Observe same precautions as for STB. Pure undiluted calcium hypochlorite will burn on contact with VX, HD, or DS2. More corrosive than STB. Will destroy clothing, has a toxic vapor, and will burn the skin. Protective mask and rubber gloves are the minimum protective equipment for handling calcium hypochlorite. Skin or clothing that comes in contact with decontaminant should be flushed with large amounts of water. Equipment that has been used to spray decontaminant must be thoroughly cleaned after the spray mission (thoroughly rinse with hot water, 80°C or 176°F).	**Chemical**—Mix 5 lb decon- taminant to 6 gal of water (10 percent solution). **Biological**—Mix 1 lb decontaminant to 6 gal water (2 percent solution). **PDDE**—Mix a slurry of 1 part decontaminant to 2 parts water (any heavier slurry will clog the decon apparatus). **Use only if STB is not Available.** A slurry of 3 parts HTH and 97 parts water can be used for horizontal surfaces. Approximate coverage is 1 gallon per 8 sq yd.
2-Propanone (acetone)	Cml	Melting point -95.35°C; boiling point 56.2°C. Evaporates rapidly. Good decontaminant for use in arctic regions. Commonly obtained as finger-nail-polish remover or paint thinners. Scrubbing increases effectiveness.	Extremely flammable. Does not neutralize agents. Effective for dissolving and flushing agent by physically removing.	None
Diethyl Ether	Cml	Melting point-116.2°C; boiling point 34.15°C. Good decontaminant for use in arctic regions. Available through medical supply facilities. Scrubbing increases effective-ness.	Same as 2-propanone.	Same as 2-propanone.
Ethylene Glycol	Cml	Contaminated surfaces should be scrubbed with decontaminat and thoroughly rinsed.	Effective in physically removing contamination, but does not neutralize the contamination. Runoff residue must be considered contaminated.	Mix 50 percent solution to 50 percent water.
Solvents (gasoline, JP-4, diesel fuel, kersoene, and similar solvents)	Cml	Contaminated surfaces should be scrubbed with decontaminate and thoroughly rinsed.	Same precautions listed for ethylene glycol are applicable to solvents. Solvents may damage materials such as rubber and plastic.	None

Table A-2.—NonStandard Decontaminants—Continued

Decontaminant	Use	Remarks	Cautions	Preparation
Sodium Hydroxide (caustic soda or lye) NSN 6810-174-6581 (100 lb)	Bio Cml	Effective against G agents, lewisite, and all biological materials including bacterial spores. Neutralized G agents on contact. Allow to remain in contact with surface contaminated with chemical agent for approximately 15 minutes.	Damaging to skin, eyes, and clothing on contact in either solution or solid form. Inhalation of the dust or concentrated mist can cause upper respiratory or lung damage. Full rubber protective clothing, gloves, boots, and mask required. In case of contact, wash area immediately with large amounts of water, flush with diluted acetic acid or vinegar. Remove affected clothing immediately. If eyes are involved, flush them at once with large amounts of warm water and get medical attention. Runoff from decon operations is highly corrosive and toxic. Drain runoff into sump and bury. All equipment should be flushed with large amounts of clear water to minimize the danger of operators being burned by residual deposits. Not recommended for ship use. Top deck storage only. The M12A1 PDDE will withstand the highly corrosive action of caustic soda solutions. However, thorough rinsing after use is necessary. Corrosive to most metals. Effectiveness is directly proportional to strength of solution. Lye is not recommended as a decontaminant if less caustic decontaminants are available. Will cause a red-color change upon contact with M8 paper.	**Small amount**—10 lb lye to 12 gal water (10 percent solution). Mix in an iron or steel container (never aluminum, zinc, or tin). Add lye to water to prevent boiling and splattering due to excessive heat emitted. Do not handle mixing container with bare hands. **Large amount**—(PDDE use) Prepare a solution of 227 grams (1/2 lb) of lye for each gal of water. Pump 350 gal of water into tank unit. Connect tank unit and heater together. Heat water to 50°C (122°F). Disconnect heater unit and add 79 kg (175 lb) of lye (1-3/4 drums) to the heated water. Circulate solution with the pump unit until all lye is disolved. The temperature will increase noticeably. Use while hot. **Simultaneous mixing and applying**—Sprinkle dry lye on the contaminated area and then dissolve it with a spray of steam or hot water. Do not wash. Do not wash the lye off the surface while applying the steam or hot water. **Paint removal**—1 lb lye per 2-1/2 gal of water is capable of removing an average coat of paint from about 11 sq yd of surface. This solution is effective in removing paint onto which chemical contamination has abosorbed. (Calcium hydroxide, potassium hydroxide, or trisodium phosphate may be substituted for sodium hydroxide).
Sodium Carbonate (washing soda, soda ash, sal soda, or laundry soda)	Cml	Effective against G agents and CN. Reacts rapidly with G agents, normally within 5 minutes. Preferred decontaminant for ship use. Recommended 5 percent by weight concentrations. No storage problem. A hot solution is the most effective means of decontaminating CN.	Do not use for VX. It can not detoxify VX and creates extremely toxic by-products. HD does not dissolve in solution and is not detoxified.	Mix 10 lb washing soda to 12 gal water (10 percent solution).
Potassium Hydroxide (caustic potash)	Bio Cml	Same remarks applicable to sodium hydroxide apply to potassium hydroxide.	Same precautions applicable to sodium hydroxide apply to potassium hydroxide.	Same preparations as used for sodium hydroxide are used for potassium hydroxide.

Decontaminant	Use	Remarks	Cautions	Preparation
Hexachloramelamine	Cml	Effective against mustard agents.	Protective mask and rubber gloves should be worn when working with hexachloramelamine. Corrosive to metal.	Decontaminant is a powder that is not soluble in water, but is soluble in organic solvents such as gasoline, kerosene, and paint thinner.
Ammonia or Ammonium Hydroxide (household ammonia)	Cml	Effective against G agents. Slower acting than sodium hydroxide or potassium hydroxide.	Self-contained breathing apparatus or special purpose mask required when working with ammonia or ammonium hydroxide.	Ammonium hydroxide is a water solution of ammonia. No further mixing is required.
Perchloroethylene (tetrachloroethylene)	Cml	Melting point -22°C; boiling point 121°C, good for use in arctic climates. A nonflammable, synthetic solvent widely used in dry cleaning plants. Dissolves H and V agents but not G. Low toxicity.	Physically dissolves and removes contamination but does not neutralize it.	No mixing required (practically insoluble in water). Effectivness increased with scrubbing.
Dichloramine-B and Dichloramine-T	Cml	Effective against mustard agents.	Protective mask and rubber gloves should be worn when working with decontaminant. Corrosive to metal.	Decontaminant is a powder that is not soluble in water but is soluble in certain organic solvents. Normally mixed as a 10 percent solution in dichlorethane.
Acids (sulfuric acid, hydrochloric acid, acetic acid, oxalic acid, and similar acids)	Nuc	Effective solvents for rust and mineral deposits holding radioactive material on metal surfaces. Normally allow 1 hour contact time. Flush with water, scrub with a water-detergent solution, flush again with water.	Acids are difficult to handle. They are harmful to the body, partiularly the eyes. Mixing acids or acid and water can produce boiling and splattering of the solution. Rubber boots, rubber gloves, rubber aprons, and goggles should be worn. Respiratory protection required in closed areas. In case of body contact with acid, flush area immediately with water. A 5 percent solution of water and baking soda (sodium bicarbonate) is used to wash acid from eyes and body. This solution neutralizes the acid.	None
Miscellaneous Solutions	Cml	To be effective, these solutions should be scrubbed onto the contaminated surfaces.	None	**MEA solution**—10 percent monoethanolamine, 1.0 percent 9N9-nonionic surfactant (triton X100) in water **WGD (C-8) solution**—302 gal water, 250 lb HTH, 36 gal perchloroethylene, and 33 lb IHF emulsifier. **ASH solution**—0.2 percent pure $Ca(OCl)_2$ from STB in water buffered to a pH of 7.53 with NaH_2 HPO_4 and .05 percent triton X100 surfactant. **Slash solution**—aqueous hypochlorite salt, aqueous citric acid, sodium citrate buffer with detergent in equal proportion (pH 7.5).

Table A-3.—Natural decontaminants.
(Decontaminants readily available and frequently occurring in nature.

Decontaminant	Use	Remarks	Cautions	Preparation
Water	Nuc Bio Cml	Flush contamination from surfaces with large amounts of water.	Effective in physically removing contamination, but does not neutralize the contamination.	None
Steam	Nuc Bio Cml	The use of steam accompanied by scrubbing is more effective than the use of steam alone.	Effective in physically removing contamination. However, contamination may not be neutralized.	None
Absorbents (earth, sawdust, ashes, rags, and similar materials)	Cml	Used to physically remove gross contamination from surfaces.	The contamination is transferred from the surface to the absorbent. The absorbent becomes contaminated and must be disposed of accordingly. Sufficient contamination to produce casualties may well remain on surfaces.	
Sealants (concrete, asphalt, earth, paint, and similar materials)	Nuc Bio Cml	Used to physically seal in or shield contamination. Various sealants are effective as follows— • 12 inches of earth provides good protection from fallout (3 inches will reduce the dose rate by about one half). • 1 inch of asphalt or concrete completely absorbs alpha and beta radiation. • 1/4 inch of grout shields alpha and beta radation. • Burying items contaminated with biological agents is an effective means of sealing off contamination. • 4 inches of earth provides good protection from chemical contamination	A break in the surface of the sealant will expose the contamination. Contaminated areas covered with sealants must be marked with appropriate NBC warning signs.	None

APPENDIX V

Decontamination of Specific Items

This appendix lists more than two dozen specific surfaces or materials, and explains briefly how to best decontaminate each for chemical, biological, or nuclear contamination. The best method of decon for a particular surface or material in a given situation could be any of those listed for that surface or material. The order in which the methods are listed does not indicate preference of one over another.

Surface or Material	Type of Contamination		
	Chemical	**Biological**	**Nuclear**
Asphalt: Roads (Applicable to small vital areas only)	• Flush with water. • Spray with slurry from PDDE. • Cover with STB; when liquid contamination is visible and personnel are nearby, use dry mix. • Weather. • Cover small areas or paths across roads with 10 cm (4 inches) of earth.	• Weather. (Remain masked). • Wet with water (will help prevent secondary aerosols, but does not decon). • Apply 2% household bleach solu tion. • Spray with slurry from PDDE. • Pour, spray, or spread oil on surface (will help prevent secondary aerosol, but does not decon).	• Brush or sweep. • Flush with water (this may drive some of the contamination into the surface; waste must be controlled). • Vacuum cleaning.
Roofs	Same as for asphalt roads.	• Same as for asphalt roads. • Apply detrochlorite (leave on at least 30 minutes, then flush with water).	Same as for asphalt roads.
Brick & Stone: Roads (Applicable to small vital areas only)	• Weather. • Spray with slurry from PDDE or apply with brushes and brooms. Let remain 24 hours, then flush with water. • Wash with soapy water, preferably hot. • Cover small areas or paths cross roads with 10 cm (4 inches) of earth.	Same as for asphalt roads.	• Same as for asphalt roads. • Abrasion (sandblasting). This provides direct and complete removal of contaminated dust; however, sand and equipment being used becomes contaminated.
Buildings	• Spray with slurry from PDDE or apply with brushes and brooms. Let remain 24 hours, then flush with water. • Use STB or dry mix around buildings where wastewater runs. • Wash with soapy water, preferably hot. • Weather.	• Same as for asphalt roads. • Apply STB slurry to vertical surfaces by manual means or PDDE. Slurry may be left on exteriors.	Same as for brick and stone roads.
Concrete: Roads (Applicable to small vital areas only)	• Spray with slurry from PDDE. • Cover with STB or dry mix. • Weather. • Cover small areas or paths across roads with 10 cm (4 inches) of earth.	Same as for asphalt roads.	Same as for brick and stone roads.

Surface or Material	Type of Contamination		
	Chemical	Biological	Nuclear
Buildings, bunkers, gun emplacements, tank obstacles	Same as for brick and stone buildings.	Same as for brick and stone buildings.	Same as for brick and stone buildings.
Earth: Roads (Applicable to small vital areas only), gun emplacements, bivouac areas, pathways, bomb craters	• Spray with slurry from PDDE. • Cover with STB; when liquid contami nation is visible and personnel are nearby, use dry mix. • Weather. • Burn (may present downwind vapor hazard). • Cover small areas or paths across roads with 10 cm (4 inches) of earth. • Scrap layer of contaminated earth to side or road.	• Same as for asphalt roads • Burn.	• Earthmoving (removal). Contaminated dust should be controlled. Equipment may become contaminated. Waste disposal must be considered. • Sealing (with earth). No waste disposal problem; however, equipment may become contaminated.
Fabrics: Canvas, covers, tarpaulins, tentage, mask carriers, web gear, clothing	Cotton • Immerse in boiling soapy water for 1 hr (1 lb soap to 10 gal water); stir. • Use 5% solution of sodium carbonate for G agents. Immerse in boiling water for 1 hr. • Launder by standard methods. • Use slurry. • Weather (except for V agents). Woolen (DS2 not recommended) Immerse in warm (100°F, soapy water for 1 hr or longer with light agitation; dry items slowly (fabric may shrink)).	Cotton • Boil in water for 15 minutes. • Autoclave for 45 minutes at 123°C (253°F). • Immerse in 2% household bleach solution for 30 minutes, rinse immediately. • Launder (destroys or inactivates all but highly resistant spores). Woolen (DS2 not recommended). Launder (fabric may shrink).	Cotton and Woolen (DS2 not recommended for woolen). • Brushing (removes contaminated dust, but presents dust hazard top personnel). • Laundering (most practical procedure; waste must be controlled; fabric may shrink).
Leather: Boots, gloves, and other items	• Scrub with hot, soapy water and rinse. • Immerse in soapy water at 120°F for 4 hrs and rinse. • Use 5% sodium carbonate solution for G agents. • Air.	• Immerse in 2% household bleach solution. Rinse. • Immerse in 2% peracetic acid for 10 minutes, rinse, and air for 10 to 50 minutes. • Wipe with 2% peracetic acid, remove excess, and air 10 to 15 minutes.	• Brushing • Flushing with water or soapy water.
Glass: Windows	• Decon kit, individual equipment. • M258A1 or M280 kit. • DS2. • Wash with hot, soapy water. • Wash with clear water or organic solvent. • Blot off surface. • Air. • Weather.	• M258A1 or M280 kit. • Wash with soap and water. • Wipe with disinfectant solution or 2% peracetic acid (see similar procedures below for mess gear).	• M258A1 or M280 kit. • Wash with detergent. • Flush with water. • Wipe with solvents.
Lenses	• M258A1 or M280 kit. • Same as for windows (DS2 may damage lens coatings). • Decon kit, individual equipment.	• M258A1 or M280 kit. • Wipe with soap and water. • Wipe with alcohol or household bleach.	• M258A1 or M280 kit. • Brush or wipe (care must be exercised to prevent scratching of lens). • Use compressed air to blow contamination from surface.

Surface or Material	Type of Contamination		
	Chemical	Biological	Nuclear
Grass and low vegetation: Fields, open terrain	• Burn. • Spray with slurry from PDDE. • Cover with STB or dry mix. • Explode drums of STB. • Clear paths through area by use of detonating cord or other detonating devices.	• Burn. • Same as for asphalt roads	Same as for earth.
Metals (unpainted): Ammunition	• Wipe with soapy water. • Wipe with organic solvent and dry. • Air.	• Wipe with soapy water. • Wipe with 2% household bleach solution. • Air.	• Brush or wipe.
Machinery	• Use DS2. • Same as for ammunition.	• Use DS2. • Wipe with 2% peracetic acid, rinse, and air for 10 to15 minutes.	• Brush or wipe. • Wash with detergent. • Flush with water.
Mess gear and canned rations	• Immerse in boiling, soapy water for 30 minutes and rinse. • Immerse in boiling water for 30 minutes. • Spray with DS2. • Wash in hot, soapy water, rinse, and air.	• Wash with soap and water, then immerse in disinfectant solution (disinfectant, chlorine, food service, or 1/3 canteen cup of household bleach per 10 gal water). • Boil in water 15 minutes. (Not effective on toxins and bacterial spores). Immerse in 5% sodium carbonate (4 lb washing soda to 10 gallons water), rinse with potable water. • Immerse in household bleach solution (11/2 gal bleach to 25 gal water) for 30 minutes, then rinse and air for 10 to 15 minutes. • Immerse in HTH solution (1/2 lb to 25 gal water) 30 minutes, then rinse. • Immerse in STB solution (1 lb to 25 gal water) for 30 minutes, then rinse. • Immerse in 2% peracetic acid for 10 minutes, rinse, and air for 10 to 15 minutes.	• Wash with soap and water, rinse. • Brush, wipe contamination from surfaces and containers.
Metals (painted): Vehicles, weapons, equipment	• DS2 (may soften paint). • Wash with hot, soapy water and rinse. • Spray with slurry from PDDE, remove from surface in 1 hour and oil surface. • Weather. • Air. • M291 kit may be used for individual weapon decon. • M280 (DKIE) decon kit, individual equipment.	• Wash with detergent and high-pressure water stream. • Apply detrochlorite. Leave on 30 minutes, then remove by washing with a stream of water. • Steam clean, using detergent. • Use household bleach solution. • Use 2% peracetic acid.	• Brush or wipe. • Wash. • Use organic solvents, caustics (not on aluminum or magnesium surfaces), complexing agents (of small value on weathered surfaces), or abrasives.

Surface or Material	Type of Contamination		
	Chemical	Biological	Nuclear
Plastics (opaque): Insulation, telephones, panel boards.	• DS2 (may soften or damage some plastics). • Wash with hot, soapy water and rinse. • Weather. • Air.	Same as for lenses	• Wash with detergents. • Flush with water. • Wipe or brush.
Plastics (transparent): Eyepieces, airplane canopies	• Wash with hot, soapy water and rinse. • Weather. • Air. • Blot off surface.	Same as for lenses	Same as for plastics (opaque).
Rubber (impermeable): Aprons, suits, and other items	• Spray with DS2 and rinse atter 30 minutes. • Immerse in hot, soapy water (just below boiling point) for 1 hour; do not agitate. Rinse with clear water and hang up to dry. • For G agents, use 10% sodium carbonate solution, rinse, and air. • Apply hot, soapy water with brushes and rinse. • Spray with slurry from PDDE. • After a few minutes, wash off with clear water.	Same as for leather.	• Brushing. • Scrubbing or flushing with water or soapy water.
Rubber (natutal and synthetic): Gloves, boots	• Spray with 10% mixture of HTH and rinse. • Immerse in slurry solution for 4 hours, rinse, and air. • Use the M291 kit in emergencies. • AIR.	Same as for leather. • Wash in warm, soapy water; rinse in clear water, and dry at room temperature. • Wipe with 2% peracetic acid; wipe off excess immediately, and air 10 to 15 minutes.	Same as for impermeable rubber.
Mask facepieces and other rubber articles coming in direct contact with the skin.	• USE the M291 kit in emergencies. • Wash with warm, soapy water, • Use decon kit, individual equipment, M280.		• Wipe or brush off. • Wipe off with water and detergent (avoid wetting mask filters).
Tires, hoses, mats, insulation.	• Spray with 10% mixture of HTH and rinse. • Apply slurry paste. Allow slurry to remain at least 30 minutes, then flush with clear water (may be left on tires). • Apply hot, soapy water. • Air. • Weather.	Use same methods used for chemical docon.	Same as for impermeable rubber.
Sand (Applicable to small vital areas only): Beaches, deserts.	• Flush with water. • Spread STB or spray slurry over surtace. • Weather. • Cover paths with roofing paper. • Scrape off 5 to 10 cm (2 to 4 inches) of contaminated top layer.	• Burn. • Wet with water (will help prevent secondary aerosols, but does not decon). • Apply 2% household bleach solution. • Apply slurry of 7 parts STB and 93 parts water (by weight). • Apply sodium hydroxide.	Same as for earth.

Surface or Material	Type of Contamination		
	Chemical	**Biological**	**Nuclear**
Undergrowth and tall grass: Meadows, jungles, forests (Applicable to small vital areas only)	• Burn (downwind vapor hazard). • Spray slurry from PDDE. • Weather. • Explode drums of STB. • Clear paths with detonating cord, bangalore torpedoes, or demolition snakes:	• Burn. • Same as for sand.	To extent possible, use same procedures as for earth.
Wood (unpainted) Buildings, vehicle bodies, boxes, crates, and similar items	• Apply slurry with PDDE, brooms, or swabs. Let slurry remain 12 to 24 hours; flush and repeat application, then flush again. • Scrub with hot, soapy water and rinse. • Weather.	• Apply detrochlorite. Leave on at least 30 minutes; flush with water. • Apply STB slurry to vertical surfaces. Slurry may be left on interiors. • Weather (sun and rain eliminate most microorganisms within one day). • Burn.	• Planning. • Wash exterior with large amounts of water (some contamination may soak into surfaces.
Wood (painted surface): (DS2 may soften paint). Buildings, boxes	• Apply slurry with PDDE ,brooms, or swabs. Let slurry remain 12 to 24 hours then rinse off with water. • Scrub with hot water and rinse. Use DS2 and rinse. • Weather.	Same as for wood buildings and boxes as previously indicated.	• Wash exterior with large amounts of water. • Wipe contamination from surface.
Water	Decon of water should only be undertaken by trained water purification personnel.	• Boil small amounts 15 minutes. • Chlorinate using chlorination kit. • Add iodine water purification tablets to small amounts of water.	• Flocculation (requires special chemicals to remove suspended matter in water). • Ion exchange (removes radioions from solution).
Food: Not canned or protected by impermeable container	Food known or suspected to be contaminated with chemical agents should not be consumed until approved by veterinary personnel.	• Boil in water 15 minutes. Cook thoroughly. • Immerse in or spray with 2% household bleach solution. (Pack aged food or food which is peeled or pared may be immersed or sprayed).	Wash or trim contamination from unpackaged food.
Food: Canned, bottled, or protected by impermeable container	See mess gear and canned rations.	See mess gear and canned rations.	See mess gear and canned rations.
Personnel	• Use M291 kit on exposed skin known or suspected to be contaminated; decon kit individual equipment, M280. • Bathe with soap and water if readily available.	• Bath with soap and hot water; decon kit individual equipment, M280. • Use the M291 kit.	• Brush or wipe from skin and hair. • Bathe with soap and hot water.

APPENDIX VI

Work/Rest Table

	Stations and Work Load				
Temperature	**1-Wash**	**2-Decon Appl.**	**3-Wait/ Interior**	**4-Rinse**	**5-Check**
	Moderate	**Heavy**	**Moderate**	**Moderate**	**Light**
Cool (<68°F) (<20°C)	60 Work 0 Rest	30 Work 30 Rest	60 Work 0 Rest	60 Work 0 Rest	60 Work 0 Rest
Warm (68°F-74°F) (20°C-24°C)	45 Work 45 Rest	20 Work 20 Rest	45 Work 45 Rest	45 Work 45 Rest	50 Work 50 Rest
Hot (77°F-84°F) (25°C-29°C)	30 Work 60 Rest	15 Work 30 Rest	30 Work 60 Rest	30 Work 60 Rest	40 Work 80 Rest
Very Hot (84°F) (29°C)	20 Work 60 Rest	10 Work 30 Rest	20 Work 60 Rest	20 Work 60 Rest	25 Work 75 Rest
When operating in temperatures above 75°F, you should consider the ability of the troops to accomplish the mission. Once the troops have reached their maximum work load for heat stress, they can not recover quickly enough to accomplish the decon mission. A viable option is to postpone the decon operation until a cooler part of the day or evening. This will reduce the heat stress load on troops and increase the probability of mission success.					

APPENDIX VII

ACRONYMS

ACE	Aviation Combat Element
AO	Area of Operations
BLT	Battalion Landing Team
BSSG	Brigade Service Support Group
CARC	Chemical agent resistant coating
CE	Command Element
CEOI	Communication Electronics Operating Instruction
COC	Combat Operations Center/Commaud
CSS	Combat Service Support
CSSE	Combat Service Support Element
CSSOC	Combat Service Support Operations Center
DAP	Decontaminating apparatus
DASC	Direct Air Support Center
DECON	Shortened form of decontamination
DED	Detailed equipment decon
DKIE	Decon kit, individual equipment
DTD	Detailed troop decon
FDC	Fire Direction Center
FMFM	Fleet Marine Force Manual
FSC	Fire Support Coodination/Coordinator
FSCC	Fire Support Coordination Center
FSSG	Force Service Support Group (USMC)
GCE	Ground Combat Element
GPM	Gallons per minute
LDS	Lightweight decontaminating system
LP	Listening Post
MAF	Marine Amphibious Force
MAGTF	Marine Air-Ground Task Force
MAU	Marine Amphibious Unit
MAW	Marine Aircraft Wing
MEB	Marine Expeditionary Brigade
MEF	Marine Expeditionary Force

MEU	Marine Expeditionary Unit
MGX	MOPP gear exchange
MOPP	Mission-oriented protective posture
MSR	Main Supply Route
MSSG	Marine amphibious unit service
OP	Observation Post
OPLAN	Operational Plan
PDDA	M12A1 power-driven decon apparatus
OPSEC	Operations Security
PDDE	Power-driven decon equipment
SD	Skin decontamination
STB	Supertropical bleach
SOP	Standing operating procedures
TOA	Table of Allowance

APPENDIX VIII

REFERENCES USED TO DEVELOP THIS TRAMAN

Chapter 1

Communications, FMFM 3-30, United States Marine Corps Combat Development Command, Quantico, VA, 1989.

MAGTF Rear Area Security, FMFM 2-6, United States Marine Corps Combat Development Command, Quantico, VA, 1991.

Marine Division, FMFM 6-1, United States Marine Corps Combat Development Command Quantico, VA, 1978.

Chapter 2

Fundamentals of Defense, Part One, INO416, United States Army, Fort Benning, GA, 1988.

Marine Rifle Company Platoon, FMFM 6-4, United States Marine Corps Combat Development Command, Quantico, VA, 1978.

Chapter 3

Marine Corps Institute, MCI 7006, United States Marine Corps Combat Development Command, Quantico, VA, 1991.

Marine Rifle Company Platoon, FMFM 6-4, United States Marine Corps Combat Development Command, Quantico, VA, 1978.

Marine Rifle Squad, FMFM 6-5, United States Marine Corps Combat Development Command, Quantico, VA, 1991.

Chapter 4

Marine Corps Institute, MCI 7005A, Marine Corps Combat Development Command, Quantico, VA.

Marine Division, FMFM 6-1, United States Marine Corps Combat Development Command, Quantico, VA, 1978.

Marine Rifle Company Platoon, FMFM 64, United States Marine Corps Combat Development Command, Quantico, VA, 1978.

Marine Rifle Squad, FMFM 6-5, United States Marine Corps Combat Development Command, Quantico, VA ,1991.

Small Unit Leader's Guide to Weather and Terrain, FMFRP 0-51, Marine Corps Combat Development Command, Quantico, VA 1990.

Chapter 5

Marine Rifle Company Platoon, FMFM 6-4, United States Marine Corps Combat Development Command, Quantico, VA, 1978.

Motor Transport, FMFM 4-9, United States Marine Corps Combat Development Command, Quantico, VA, 1992.

Chapter 6

NBC Decontamination, FM 3-5, Headquarters Department of the Army, United States Marine Corps, Washington, DC, 1993.

INDEX

T

☆ U.S. GOVERNMENT PRINTING OFFICE:1996-733-127/40014

Assignment Questions

Information: The text pages that you are to study are provided at the beginning of the assignment questions.

ASSIGNMENT 1

extbook Assignment: "Organization and Operation of the Combat Operation Center," pages 1-1 through 1-16.

Learning Objective: Identify and briefly describe commands that a Seabee Unit can be attached to during combat operations.

1-1. During combat operations Seabee units are normally assigned to which of the following commands?

1. Army Engineer Corp
2. Marine Air-Ground Task Force
3. Air Force Red Horse Team
4. An independent command

1-2. A MAGTF consists of a minimum of how many elements?

1. One
2. Two
3. Three
4. Four

1-3. A MAGTF consists of a minimum of how many primary engineer assets?

1. Seven
2. Six
3. Five
4. Four

1-4. What function does the Command Element serve within a MAGTF?

1. As a supply and logistics center
2. As an intelligence center
3. As a headquarters
4. As a communication center

1-5. The Command Element within a MAGTF is established for which of the following reasons?

1. To provide effective communication
2. To provide effective planning and execution of the MAGTF elements
3. To provide supplies to the MAGTF elements
4. To duplicate the efforts of the other MAGTF elements

1-6. What is the primary duty of the CEC officer attached to the Command Element?

1. Advisor to the MAGTF supply officer on matters relating to the capabilities and employment of the NCF
2. Advisor to the MAGTF commander on matters relating to the capabilities and employment of the NCF
3. Advisor to the MAGTF communication officer on matters relating to the capabilities and employment of the NCF
4. Advisor to the MAGTF engineer officer on matters relating to the capabilities and employment of the NCF

1-7. The Ground Combat Element (GCE) is task-organized for which of the following operations?

1. Ground maneuver operations
2. Tactical air operations
3. Combat service support operations
4. Command and control operations

1-8. The GCE is constructed around what type of unit?

1. A communication unit
2. A supply unit
3. A combat infantry unit
4. A headquarters unit

1-9. The Aviation Combat Element (ACE) is task-organized to conduct which of the following operations?

1. Tactical air
2. Combat service support
3. Ground maneuver
4. Command and control

1-10. When a MAGTF commander assumes responsibility for control of air operations, he or she will normally appoint the commander of what MAGTF element as the MAGTF tactical air commander (TAC)?

1. The Combat Service Support Element commander
2. The Command Element commander
3. The Aviation Combat Element commander
4. The Ground Combat Element commander

1-11. Depending on the assigned mission, the Combat Service Support Element (CSSE) is task-organized to provide which of the following functions?

1. Engineering, supply, and transportation
2. Weapons, maintenance, and disbursing
3. Automated data processing and personnel services
4. All of the above

1-12. Normally, the Naval Construction Force is classified as what type of organization within a MAGTF?

1. Combat service support
2. Engineering service support
3. Ground combat support
4. A separate task organization

1-13. When assigned to a MAGTF, the NCF is usually OPCON to what element?

1. The Combat Service Support Element
2. The Ground Service Support Element
3. The Command Element
4. The Aviation Combat Element

1-14. There are a total of how many types of MAGTFs?

1. One
2. Two
3. Three
4. Four

1-15. Normally, what is the rank of the commander of a Marine Expeditionary Unit (MEU)?

1. Colonel
2. General
3. Captain
4. First lieutenant

1-16. Under normal circumstances, the Ground Combat Element within a MEU consists of a total of how many battalion landing teams?

1. One
2. Two
3. Three
4. Four

1-17. Under normal circumstances, the Aviation Combat Element within a MEU is what type of squadron?

1. A strategic attack
2. A observation
3. A reconnaissance
4. A helicopter

1-18. Within a MEU, what is the Combat Service Support Element normally called?

1. The FSSG
2. The MSSG
3. The CSSE
4. The MAW

1-19. Normally, what is the rank of the commander of a Marine Expeditionary Force-Forward (MEF-F)?

1. General
2. Colonel
3. Admiral
4. Brigadier general

1-20. Under normal circumstances, the Ground Combat Element within a MEF-F consists of what type of organization?

1. A regimental landing team (RLT)
2. A battalion landing team (BLT)
3. An infantry division
4. An infantry battalion

1-21. What element, if any, is the Aviation Combat Element within a MEF-F?

1. Marine Air Wing (MAW)
2. Marine Helicopter Squadron
3. Marine Air Group (MAG)
4. None

1-22. What is the Combat Service Support Element within a MEF-F?

1. MAW
2. MAG
3. BSSG
4. CSSE

23. Normally, what is the rank of the commander of a Marine Expeditionary Force (MEF)?

1. Either a major general or lieutenant general
2. Either a brigadier general or lieutenant general
3. Either a general or lieutenant general
4. Either a colonel or lieutenant

24. What is the primary function of the CEC officer attached to the Command Element of a MEF?

1. Liaison between the MEF commander and NCF units
2. Liaison between the Command Element and the Combat Service Support Element
3. Liaison between the engineering staff and NCF units
4. Liaison between the communication staff and NCF units

25. The Ground Combat Element within a MEF is a unit of what size?

1. A Marine infantry battalion
2. A Marine division
3. A Marine company
4. A Marine platoon

-26. What is the abbreviation for the ACE within a MEF?

1. MAW
2. MAG
3. MAS
4. MAD

-27. What is the abbreviation for the CSSE within a MEF?

1. CSSG
2. BSSG
3. MSSG
4. FSSG

Learning Objective: Identify and briefly describe the rear area security organization.

1-28. Seabees should be familiar with the rear area organization for what reason?

1. Seabee units are primarily assigned to the rear area
2. Seabees units are the only units in the rear area
3. Seabee units are always the main organization in the rear area
4. No need, Seabee units are never in the rear area

1-29. Today, what factor makes the rear area more vulnerable to enemy attack than in the past?

1. The enemy is smarter
2. The enemy numbers have increased
3. The modern weaponry
4. Better intelligence is available to the enemy

1-30. What does the acronym RAS stand for?

1. Rear area security
2. Rear area station
3. Rear area substation
4. Rear area supply

1-31. During combat operations, Seabee units are assigned primarily to what area?

1. Forward edge of the battle area (FEBA)
2. Main supply routes (MSR)
3. Rear area
4. Reserve area

1-32. The MAGTF rear area extends rearward from what boundary?

1. The rear boundary of the CSSE
2. The rear boundary of the GCE
3. The forward boundary of the CSSE
4. The forward boundary of the GCE

1-33. What type of defense concept requires base facilities be established in the rear area?

1. All-around
2. Perimeter
3. Cluster
4. Strategic

1-34. The rear area security coordinator (RASC) monitors operations of the rear area through what two organizations?

1. Combat Service Support Operations Center (CSSOC) and the Rear Area Operations Center (RAOC)
2. Combat Operation Center (COC) and the Fire Direction Center (FDC)
3. Fire Support Center (FSC) and the Base Facility Commander (BFC)
4. Combat Operation Center (COC) and the Base Facility Commander (BFC)

1-35. The Force Service Support Group (FSSG) can field a total of how many provisional mobile security platoons (PMSP)?

1. One
2. Two
3. Three
4. Four

Learning Objective: Identify the importance of the COC and the location of a COC.

1-36. A commander establishes a combat operation center for what primary purpose?

1. To ensure medical facilities
2. To ensure proper galley facilities
3. To ensure proper base facilities
4. To ensure proper centralized command and control facilities

1-37. What officer exercises staff supervision over the COC?

1. S-1
2. S-2
3. S-3
4. S-4

1-38. When you plan the establishment of a COC, what special consideration must be considered?

1. Location
2. Size
3. Terrain
4. Accessibility

1-39. Around what unit is the COC centrally located?

1. Platoon position
2. Galley
3. Battalion aid station
4. Company command posts

1-40. The COC should have at least how many dismount and entry points?

1. One dismount, one entry
2. One dismount, two entries
3. Two dismounts, two entries
4. Two dismounts, one entry

1-41. What is the most critical factor a COC must have in order to be successful?

1. Proper location
2. Communication
3. Light discipline
4. Security

Learning Objective: Describe the organization and mission of the Combat Operation Center (COC) and the duties of its personnel.

A. Intelligence Watch Officer
B. COC Watch Officer
C. Communication Watch Officer
D. Communicators
E. Messenger/Security
F. Fire Support Coordinator
G. NBC/CBR Officer

Figure 1A.—COC personnel.

IN ANSWERING QUESTIONS 1-42 THROUGH 1-48, SELECT FROM FIGURE 1A THE PERSON RESPONSIBLE FOR THE TASK.

1-42. Maintains close communication with the battalion's Fire Direction Center.

1. G
2. F
3. E
4. D

1-43. Ensures incoming and outgoing messages adhere to established routing procedures.

1. A
2. B
3. C
4. D

1-44. Has releasing authority on all outgoing messages.

1. D
2. C
3. B
4. A

1-45. Ensures all personnel entering the COC are on the access list.

1. B
2. C
3. D
4. E

1-46. Prepares written and oral briefings as required.

1. A
2. B
3. C
4. D

1-47. Drafting messages.

1. A
2. B
3. C
4. D

1-48. Monitors all incoming messages related to the weather.

1. D
2. E
3. F
4. G

1-49. Which of the following information is NOT likely to be posted on the operations map located in the COC?

1. Location and status of patrols
2. Landing zone locations
3. Enemy equipment/weapons
4. NBC corridors/fallouts

1-50. What organization is the terminating point for all tactical and nontactical radio nets?

1. Combat Operation Center
2. Company command post
3. Battalion communication center
4. Antenna farm

1-51. The AN/PRC-104 has what type of frequency?

1. Low
2. Variable
3. Medium
4. High

1-52. The AN/GRA-39 is primarily used for what purpose?

1. Remote linkup from the "antenna farm" to the COC
2. Remote linkup from the COC to the company command posts
3. Remote linkup from the Fire Direction Center (FDC) to the company command post
4. Remote linkup from the "antenna farm" to the company command post

1-53. What does the acronym CEOI stand for?

1. Combat Electronic Operating Instruction
2. Command Electronic Operating Instruction
3. Communication Electronic Operating Instruction
4. Combat Equipment Operating Instruction

A.	Rear Area Operations Command Net
B.	Alert/Broadcast Net
C.	Base Facility Tactical Net
D.	Conduct of Fire Net

Figure 1B.—Various radio nets.

IN ANSWERING QUESTIONS 1-54 THROUGH 1-57, SELECT FROM FIGURE 1B THE RADIO NET ASSOCIATED WITH THE TASK.

1-54. Communication from the COC to the battalion's FDC.

1. D
2. C
3. B
4. A

1-55. Communication from the COC to higher headquarters located outside the base facility, usually by a high-frequency radio.

1. C
2. D
3. A
4. B

1-56. Used to pass general information about all (or the majority) of the units within the area of operation.

1. A
2. B
3. C
4. D

1-57. Used by the COC for tactical purposes and for various reports such as SALUTE reports.

1. A
2. B
3. C
4. D

1-58. All incoming radio traffic to the COC should be written down on authorized message pads with an original plus how many copies?

1. One
2. Two
3. Three
4. Four

1-59. As a guide, a priority message should be handled within what time frame?

1. 30 minutes-1 hour
2. 1-3 hours
3. 3-4 hours
4. As fast as possible

1-60. After an incoming message has been received and checked by the communication chief, to whom does the communication chief pass the message to?

1. The appropriate action section on the message
2. The operations officer
3. The intelligence officer
4. The COC watch officer

ASSIGNMENT 2

Textbook Assignment: "Organization and Operation of the Company Command Post" and "Setup and Control of Medical Evacuation," pages 2-1 through 3-4.

Learning Objective: State the purpose of the company command post (CP) and what must be considered when establishing a company command post.

2-1. In support of the company commander, the company command post (CP) serves what primary function?

1. Communication to the platoons
2. Centralized command and control
3. A place for the platoon commanders to meet
4. Planning purposes

2-2. In support of the COC, the company command post serves what primary function?

1. It is the "eyes and ears" for the COC
2. It defends the COC against attacks
3. It assists the COC in making decisions relating to combat operations
4. It assists the COC in project planning

2-3. You should establish alternate routes leading to the company command post for which of the following reasons?

1. To prevent personnel from getting lost trying to find the CP
2. So communication to the platoons will not be lost
3. To ensure personnel will be able to establish a sector of fire
4. To prevent a beaten path for the enemy to follow from being established

2-4. Which of the following elements is NOT included in the company CP's communication network?

1. The react force
2. The platoon commanders
3. The listening post
4. The observation post

2-5. The platoon commanders must have constant communication with which of the following elements/personnel?

1. The COC
2. The react force
3. The troops on the front lines
4. The other companies

2-6. A break in communication can cause mass confusion and could result in the loss of lives.

1. True
2. False

2-7. Which of the following factors is a principal concern to the company commander?

1. Chow for the troops
2. Defending the company's area of responsibility
3. Communication to the COC
4. Defending the COC

2-8. The company CP is centralized around what command element to enhance command and control?

1. The platoons
2. The COC
3. The galley
4. The battalion aid station

Learning Objective: Describe the organization of the company command post.

2-9. Standard Operating Procedures (SOP) should be established in the company CP for which of the following reasons?

1. To hold people accountable for their actions
2. To keep the COC informed
3. To avoid confusion
4. To keep the commanding officer informed

2-10. The watch chief should perform which of the following actions before assuming watch?

1. Ensure the platoons are all accounted for
2. Read the message board
3. Ensure all watch personnel are present
4. Make liaison with the COC

2-11. Operational requirements can be accomplished when what number of personnel are assigned to watch stations in the company CP?

1. 1
2. 2
3. Minimum
4. Maximum

A.	Watch Chief
B.	Watch Petty Officer
C.	Messenger/Security
D.	Communicator

Figure 2A.—Watch station personnel.

IN ANSWERING QUESTIONS 2-12 THROUGH 2-14, SELECT FROM FIGURE 2A THE PERSON RESPONSIBLE FOR THE TASK.

2-12. Handle and process information relating to intelligence.

1. A
2. B
3. C
4. D

2-13. Ensures all incoming and outgoing messages adhere to establish procedures.

1. A
2. B
3. C
4. D

2-14. Ensures only authorized personnel enter the company CP.

1. D
2. C
3. B
4. A

2-15. Normally, there is a total of how many maps in the company CP?

1. One
2. Two
3. Three
4. Four

2-16. What is the primary reason charts are posted in the company CP?

1. To post the daily password
2. To post the plan of the day
3. To post a quick picture of the current situation
4. To post the company watch bill

2-17. From the information listed below, select the one that is NOT usually posted on a chart.

1. Company strength
2. Convoy status
3. Call signs
4. Company fire plan

Learning Objective: Describe the contents of a communication plan.

2-18. A communication plan includes which of the following information?

1. Communication network
2. Company fire plan
3. Battalion fire plan
4. Platoon fire plan

2-19. From the types of communication equipment listed below, select the one that would be ideal for communication from the platoon commander's fighting position to the squad leader's fighting position.

1. SB-22
2. AN/PRC-199A
3. TA-312
4. GRA-29

2-20. What officer is responsible to the commanding officer for the unit's communication plan?

1. The company commander
2. The communication officer
3. The operations officer
4. The executive officer

2-21. The radio nets that companies are required to monitor are posted in what instruction?

1. SOP
2. 15500/1
3. 5000 series
4. CEOI

2-22. Which of the following information is a primary factor to consider when designing the company's radio and wire plans?

1. The number of personnel in the company
2. The number of fighting positions
3. Communication assets available to the company
4. The company's fire plan

2-23. Normally, there is a total of how many frequency nets a company is required to monitor?

1. One
2. Two
3. Three
4. Four

Learning Objective: Describe the operations of the company command post.

2-24. In the company CP, all incoming and outgoing messages should be written on authorized message pads to include the original plus how many copies?

1. One
2. Two
3. Three
4. Four

A.	Watch Chief
B.	Watch Petty Officer
C.	Messenger/Security
D.	Communicator

Figure 2B.—Watch station personnel.

IN ANSWERING QUESTIONS 2-25 THROUGH 2-29, SELECT FROM FIGURE 2B THE PERSON RESPONSIBLE FOR THE TASK.

2-25. Initially receives an incoming message by radio.

1. A
2. B
3. C
4. D

2-26. Initially checks incoming messages for format.

1. A
2. B
3. C
4. D

2-27. Prepares outgoing messages.

1. A
2. B
3. C
4. D

2-28. Decides on the appropriate action to take for incoming messages.

1. A
2. B
3. C
4. D

2-29. Files a copy of outgoing messages before transmission.

1. A
2. B
3. C
4. D

2-30. What are the two most common reports used by the companies to relay information quickly?

1. METT-T and SITREP reports
2. NBC and CASREP reports
3. SALUTE and SPOT reports
4. WIA and ZULA reports

2-31. What type of communication is used for primarily communicating between the platoon commander's fighting position and the squad leader's fighting position?

1. Wire
2. Radio
3. Hand signals
4. Messenger

2-32. You are a platoon commander and your platoon has engaged in a fire fight. What type of report, if any, should you immediately send to the company CP?

1. SALUTE report
2. WIA report
3. SPOT report
4. None

2-33. You observe enemy activity but it does not pose a threat to the company. You should send immediately what type of report to the company CP?

1. SALUTE report
2. WIA report
3. SPOT report
4. NBC report

2-34. Which of the following personnel maintains the situation map located in the company CP?

1. The watch petty officer
2. The messenger
3. The communicator
4. The watch chief

2-35. What responsibility is the primary duty of the company CP?

1. Keeping the COC informed
2. Defending the company's area of responsibility
3. Keeping the troops informed
4. Maintaining liaison with the other companies

2-36. A SPOT report is followed up with what other type of report?

1. WIA report
2. NBC report
3. SALUTE report
4. SITREP report

2-37. What critical support does the COC depend on from the company CP?

1. Ammunition
2. Communication personnel
3. Accurate and concise information
4. Security personnel

Learning Objective: Describe in detail the setup, the control, the preparation, and the difference between landing zones, landing sites, and landing points.

A. Landing Zone

B. Landing Site

C. Landing Point

D. Landing Station

Figure 2C.—Components of a landing zone.

IN ANSWERING QUESTIONS 2-38 THROUGH 2-40 REFER TO FIGURE 2C.

2-38. Identified by color.

1. A
2. B
3. C
4. D

2-39. Identified by a code name.

1. A
2. B
3. C
4. D

2-40. Identified by a two-digit number.

1. A
2. B
3. C
4. D

2-41. What factors, if any, determines how landing zones are divided?

1. By landing points
2. By landing sites
3. By terrain
4. None

2-41. Landing points are primarily assigned to what location?

1. Within landing sites
2. Within landing zones
3. Within landing sections
4. Within LZs

2-43. What data is the first factor that must be considered when you are preparing a landing zone?

1. Terrain
2. Equipment
3. Type of helicopter using the landing zone
4. Defense of the landing zone

2-44. When you are preparing a LZ, approaches and exits must be cleared of obstacles that are of what minimum height?

1. 5 meters and higher
2. 10 meters and higher
3. 15 meters and higher
4. 20 meters and higher

2-45. When you are preparing a LZ, approaches and exits must be cleared of obstacles extending at least how many meters in the direction of the approach and exit path?

1. 20
2. 30
3. 40
4. 50

2-46. When determining the distance between a landing point and a high obstruction, you should use what "rule of thumb" ratio?

1. 10:1
2. 5:1
3. 15:1
4. 20:1

2-47. Using the rule of thumb ratio, a helicopter landing or taking off near a 40-foot tree needs at least how many feet of horizontal clearance?

1. 200
2. 300
3. 400
4. 500

2-48. When you are preparing a landing zone, obstacles, such as rocks and stumps, should not exceed what height?

1. 1 foot
2. 2 feet
3. 3 feet
4. 4 feet

2-49. In order for helicopters to land safely, landing points should not have a ground slope exceeding how many degrees?

1. 10
2. 12
3. 14
4. 16

2-50. In what direction do helicopter pilots prefer for landing on a slope?

1. Parallel to the slope
2. Perpendicular to the slope
3. Downhill
4. Uphill

2-51. Any river or stream which has a firm bottom and is no deeper then 20 inches can be used as a landing site.

1. True
2. False

2-52. A helicopter is required to land into the wind when the ground level wind exceeds how many knots?

1. 3 knots
2. 5 knots
3. 10 knots
4. 15 knots

2-53. When you are using the grass drop method to determine wind velocity, what number is the angle formed between your arm and your body divided by?

1. 5
2. 2
3. 3
4. 4

2-54. You use the angle of the smoke method to determine the wind velocity. What is the velocity of the wind if the smoke blows at a 30-degree angle?

1. 2-4 knots
2. 3-5 knots
3. 4-6 knots
4. 5-7 knots

2-55. At what angle will smoke blow when the wind velocity is between 5-7 knots?

1. 30 degrees
2. 40 degrees
3. 50 degrees
4. 60 degrees

2-56. When marking a landing zone during the day, you should use what color for smoke and landing zone panels?

1. Same color as the landing zone designation
2. Opposite colors
3. Always use red
4. Same color as the landing site designation

2-57. When using smoke to mark a landing zone, you should never tell the pilot the color of the smoke.

1. True
2. False

2-58. When using the vector instruction method, you use what device to guide the aircraft to the landing zone?

1. A flag
2. A smoke grenade
3. A radio
4. Landing panels

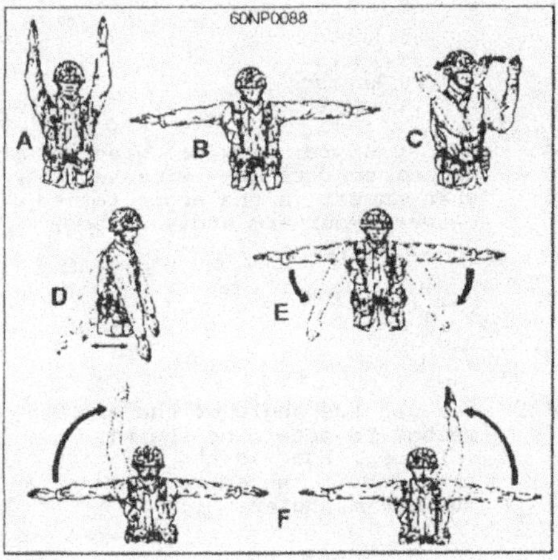

Figure 2D.—LSE signals.

IN ANSWERING QUESTIONS 2-59 THROUGH 2-64, SIGNAL.

2-59. LSE signal for guiding a helicopter downward.

1. E
2. B
3. F
4. D

2-60. LSE signal for guiding a helicopter forward.

1. A
2. B
3. C
4. D

2-61. LSE signal used to hover a helicopter above the ground.

1. F
2. E
3. D
4. B

2-62. The LSE signal used for informing the helicopter pilot to prepare for guidance.

1. A
2. B
3. C
4. D

2-63. The LSE signal used for guiding a helicopter right or left.

1. D
2. E
3. F
4. A

2-64. The LSE signal used for informing the pilot to move the helicopter back.

1. D
2. A
3. E
4. B

ASSIGNMENT 3

extbook Assignment: "Setup and Control of Medical Evacuation" and "Planning and Development of Defense Tactics," pages 3-6 through 4-8.

Learning Objective: Describe the contents of a landing zone brief.

3-1. Normally, a landing zone brief is broken down into a total of how many lines?

1. Fourteen
2. Sixteen
3. Twelve
4. Ten

3-2. What information is contained in line 2 of a landing zone brief?

1. Call signs
2. Elevation of the landing zone
3. Obstacles in the landing zone
4. Location of a landing zone

3-3. What information is contained in line 6 of a landing zone brief?

1. Mission number
2. Wind direction
3. Call signs
4. Location of a landing zone

3-4. What information is contained in line 9 of a landing zone brief?

1. Your call sign
2. The pilots call sign
3. Friendly positions
4. Enemy positions

3-5. What information is contained in line 10 of a landing zone brief?

1. Approach heading
2. Location of friendly forces
3. Location of enemy forces
4. Obstacles

Learning Objective: Describe the contents of a MEDEVAC request and MEDEVAC procedures.

3-6. What command element/unit makes the final decision concerning a MEDEVAC request?

1. The DASC
2. The COC
3. The RAOC
4. The BAS

3-7. As a platoon commander, what action should you take when you need a MEDEVAC but are unable to communicate with the COC, the BAS, or the company CP?

1. Send a messenger to the CP
2. Send a messenger to the COC
3. Send a messenger to the BAS
4. Route a MEDEVAC request to the next higher authority

3-8. On a MEDEVAC request, what information is documented on line 9?

1. Landing zone call sign
2. Requesting unit call sign
3. Landing zone coordinates
4. Landing zone markings

3-9. On a MEDEVAC request, what information is documented on line 4?

1. Date-time group
2. Number of personnel needing MEDEVAC
3. Landing zone location
4. Airborne medical assistance required

3-10. Under an ideal situation, litter bearers should be personnel from what company?

1. Headquarters Company
2. Alpha Company
3. Bravo Company
4. Delta Company

3-11. Normally, what total number of litter bearers per company is considered adequate?

1. Six
2. Eight
3. Ten
4. Four

3-12. After administrating basic first aid to a critically wounded person in your company, you should take what course of action next?

1. Shout for a corpsman
2. Quietly pass the word to the company CP that a corpsman is needed
3. Send a messenger to the COC
4. Send a messenger to the battalion aid station

3-13. What is the primary function of an ambulance?

1. To give first aid
2. To prevent the prolong carrying of litters
3. To provide medical supplies for the companies
4. To acquire medical supplies for the battalion

3-14. Prior to a medical evacuation, casualties are brought to what central location?

1. The company CP
2. The COC
3. The landing zone
4. The battalion aid station

3-15. What person has the final say on whether a casualty needs to be evacuated?

1. A member of the medical staff
2. The COC
3. The company commander
4. The operations officer

3-16. What title is given to the person on the security team that controls all movements of the wounded to the landing zone?

1. "Pit Boss"
2. Landing zone coordinator
3. Security team leader
4. React force team leader

3-17. What factor dictates the weapons and personnel required to secure a landing zone?

1. The weapons available in the TOA
2. The number of personnel assigned to the security team
3. The size and terrain of the landing zone
4. The number of casualties being evacuated

3-18. A total of how many radios are required for communication in a landing zone?

1. One
2. Two
3. Three
4. Four

3-19. How are casualties transported from the battalion aid station to the landing point?

1. By ambulance
2. By litters
3. By any means available
4. By the security team

3-20. Before the arrival of a helicopter, which of the following is a primary way casualties are arranged at the landing point?

1. By the capacity of the helicopter
2. By rank
3. By fire teams
4. By squads

3-21. What person is responsible for directing the helicopter representative to the senior medical staff personnel in charge?

1. The security team leader
2. A representative from the COC
3. The landing signalman
4. The operations officer

3-22. What is the primary reason helicopter evacuation should be used with discretion?

1. Helicopters draw enemy fire
2. A helicopter evacuation takes a lot of coordination
3. A helicopter evacuation removes personnel from the defensive perimeter
4. A helicopter evacuation requires removing some weapons from the defensive perimeter

---

Learning Objective: State the importance and purpose of defensive combat.

3-23. Normally, the battalion security area is located at what distance in front of the defensive perimeter?

1. 200 meters
2. 300 meters
3. 400 meters
4. 500 meters

3-24. Which of the following elements are NOT a Seabee force located in the security area?

1. Listening posts
2. Observation posts
3. Patrols
4. Machine gun teams

3-25. The forward defense area extends rearward from the defensive lines to the rear boundary of what command organization?

1. The company CP
2. The crew-served weapons
3. The frontline companies
4. To the rear of the security area

3-26. The composition of the forward defensive echelon is planned according to what concept?

1. The terrain
2. The type of defense
3. The number of personnel
4. The number of weapons in the TOA

3-27. The reserve force serves what purpose in the overall defensive plan?

1. To keep the frontline companies supplied with ammo
2. To man the AT4s for the front line
3. To man crew-served weapons flanking the front lines
4. To add depth to the defense

Learning Objective: Describe the principles of the fundamentals of defense.

3-28. What factor virtually decides the positioning of platoons in the defense?

1. Weapons
2. Personnel
3. Equipment
4. Terrain

3-29. The analysis of weather and terrain is known by what acronym?

1. METT-T
2. BAMUS
3. KOCOA
4. FLOT

3-30. In comparison to aviation units, service support units require what type of terrain for combat operations?

1. Suitable terrain to construct roads for moving supplies
2. High terrain
3. Low terrain
4. Dense terrain

3-31. When you are planning the defense, what consideration is so closely related to fields of fire that they are considered together?

1. Observation
2. Camouflage and concealment
3. Avenues of approach
4. Communication

3-32. When establishing fields of fire for a weapon, you MUST KNOW what information about the weapon?

1. The maximum effective range of the weapon
2. The maximum range of the weapon
3. The capabilities of the weapon
4. The sustain rate of fire of the weapon

3-33. When engaging the enemy within the field of fire of a M60E machine gun, you should adhere to which of the following rules?

1. Engage the enemy when they reach established barriers
2. Engage the enemy as far out as possible
3. Engage the enemy close enough so the position of the machine gun is not compromised
4. Engage the enemy at the forward edge of the security area

3-34. Of the items listed below, which one is classified as cover and concealment?

1. Rocks and shell craters
2. River banks and underbrush
3. Woods and snowdrifts
4. Ditches and quarries

3-35. Entanglements are primarily used for what purpose?

1. To prevent the enemy from moving forward
2. To stop the enemy in their tracks
3. To channel the enemy away from the defensive perimeter
4. To channel the enemy into a beaten zone

3-36. What is the primary factor that HAMPERS the achievement of mutual fire support?

1. A well-developed defense plan
2. A well-developed fire plan
3. The widespread separation of fire teams
4. The widespread separation of units

3-37. A convoy has traveled out or the range of their unit's organic weapons. What action is taken to ensure the convoy is protected?

1. A convoy should never travel out of range from their unit's organic weapons
2. Before departing, the convoy commander coordinates for mutual support from adjacent units
3. Before departing, the convoy commander ensures there are enough weapons to defend the convoy independently
4. Before departing, the convoy commander informs the COC of the areas the convoy will be out of range from the unit's organic weapons

3-38. What direction from his position must a platoon commander be prepared to defend against?

1. All directions
2. The excepted direction of an enemy attack
3. Depends on the terrain
4. Depends on the weapons assigned to the platoon

3-39. When a Seabee unit is attached to a MAGTF, what element provides early warning of airborne attacks?

1. MEB
2. CSSE
3. MAW
4. GCE

3-40. Of the items listed below, which one is NOT included in a coordinated fire plan?

1. Naval gunfire
2. The battalion's mortar crew
3. Tactical aviation support
4. Anti-personnel mines

Learning Objective: Organize and position a squad in the defense.

3-41. What primarily influences the dispersion of squads in the defense?

1. The terrain
2. The mission
3. Enemy nuclear capability
4. Mobility of opposing forces

3-42. In a rifle company, what person ha the primary duty for positioning squads in the defense?

1. The company commander
2. The platoon commander
3. The battalion commander
4. The operations officer

3-43. Compared to infantry units, mechanized units use what type of terrain to attack?

1. Rivers
2. Forest
3. Fields
4. Hills

3-44. "Seek to fire on the enemy when he cannot fire on you and do not let the enemy fire on you when you cannot fire on him."

1. True
2. False

3-45. Which of the following sectors of fire, if any, is assigned to a squad leader?

1. An individual sector of fire
2. A sector of fire that overlaps a fire team
3. A sector of fire covering a flank of a fire team
4. None

3-46. You should examine an assigned sector of fire from what position?

1. The sitting position
2. The standing position
3. The prone position
4. The freehand position

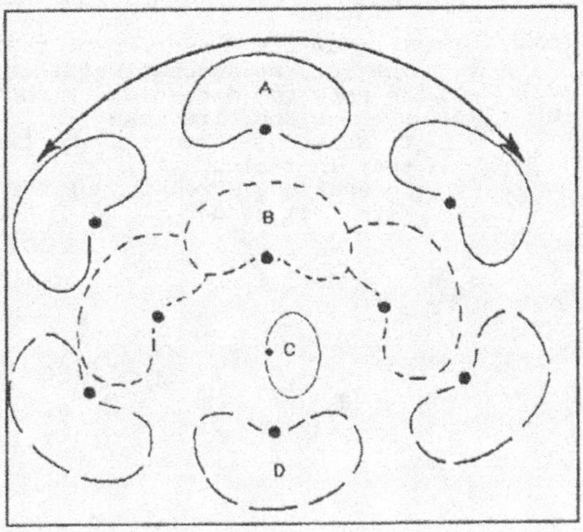

Figure 3A.—Defensive firing positions.

IN ANSWERING QUESTIONS 3-47 THROUGH 3-48, REFER TO FIGURE 3A.

3-47. What defensive firing position is the alternate position?

1. D
2. C
3. B
4. A

3-48. What defensive firing position is the supplementary position?

1. D
2. C
3. B
4. A

3-49. Which of the following elements is assigned to a alternate firing position?

1. Individual
2. Squads
3. Fire teams
4. Crew-served weapons

3-50. What is the sector of fire for the alternate firing position?

1. The supplementary position's sector of fire
2. The primary position's sector of fire
3. Setup as required
4. Depends on the enemies movement

3-51. What is the firing mission for a supplementary position?

1. To protect primary positions
2. To protect alternate positions
3. To protect against an attack from an unexpected direction
4. A backup in case the primary positions are neutralized

3-52. A total of how many PDFs may be assigned to a weapon?

1. One
2. Two
3. Three
4. Four

Learning Objective: Organize and position a fire team in the defense.

3-53. What factor is used by the squad leader and the fire team leader to show forward and lateral limits?

1. Aiming stakes
2. Terrain features
3. T&E readings
4. Grid north

3-54. Within a fire team, what distance is used as the interval between fighting positions?

1. 5-20 meters
2. 10-25 meters
3. 15-25 meters
4. 15-30 meters

3-55. What distance is used as the interval between single-fighting positions?

1. 5 meters
2. 10 meters
3. 15 meters
4. 20 meters

3-56. What distance is used as the intervals between double-fighting positions in open terrain?

1. 5 meters
2. 10 meters
3. 15 meters
4. 20 meters

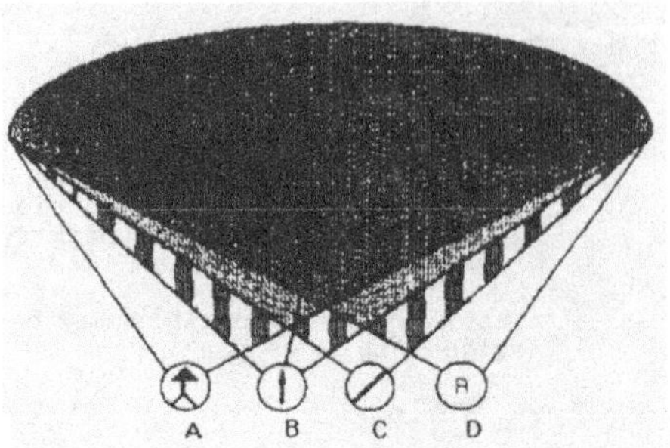

Figure 3B.—Fire team fire plan.

IN ANSWERING QUESTIONS 3-57 THROUGH 3-60, REFER TO FIGURE 3B.

3-57. What letter identifies the symbol for a rifleman?

1. A
2. B
3. C
4. D

3-58. What letter identifies the symbol for the automatic rifleman?

1. A
2. B
3. C
4. D

3-59. What letter identifies the symbol for the fire team leader?

1. A
2. B
3. C
4. D

3-60. What letter identifies the symbol for the assistance automatic rifleman?

1. A
2. B
3. C
4. D

3-61. At what location is the fire team leaders positioned within the fire team?

1. To the left of the fire team
2. To the right of the fire team
3. In the center of the fire team
4. In back of the fire team

3-62. What is the purpose of the magnetic north line on a fire plan sketch?

1. To show the direction of the PDF for the automatic rifleman
2. To show the direction of the FPL for the fire team
3. To show the direction the fire team is facing
4. To show the direction of the fire team leader

ASSIGNMENT 4

'extbook Assignment: "Planning and Development of Defensive Tactics" and "Counter Ambush Techniques," pages 4-8 through 5-5.

Learning Objective: Organize and position a squad in the defense.

4-1. What person assigns each squad a defensive mission?

1. The company commander
2. The platoon commander
3. The squad leader
4. The company chief

4-2. Which of the following factors dictates the effective placement of fire teams in the defense?

1. Number of personnel in the fire team
2. Number of crew-served weapons assigned to the platoon
3. Number of M16s assigned to the squad
4. Terrain

4-3. Which of the following weapons are positioned before the positioning of squads?

1. Mortars
2. AT4s
3. Crew-served weapons
4. M203s

4-4. Normally, what person designates the general firing position for a specific automatic weapon?

1. Squad leader
2. Platoon commander
3. Fire team leader
4. Company commander

4-5. What is the position of the squad leader in relation to the fire team?

1. To the rear and center of the fire teams
2. To the rear and left of the fire teams
3. To the rear and right of the fire teams
4. In front of the fire teams

4-6. Which of the following positions is NOT included in a squad fire plan sketch?

1. Position of the fire teams
2. Position of on-call targets
3. PDF for automatic rifles
4. Position of the company command post

Learning Objective: Organize and position a platoon in the defense.

4-7. What officer assigns a frontage to a platoon?

1. Commanding officer
2. Company commander
3. Operations officer
4. Intelligence officer

4-8. Other than the size of the company battle area, what other factor dictates the frontage assigned to a platoon?

1. Personnel
2. Weapons
3. Terrain
4. Weather

4-9. The depth of a platoon defense is the distance between the squad's primary positions and the reward extension of what other position?

1. Crew-served weapon positions
2. Supplementary positions
3. Alternate positions
4. The company command post position

4-10. Which of the following areas is important for the platoon commander to know when a patrol is entering that area?

1. Security area
2. Defensive area
3. Forward defense area
4. Reserve area

4-11. Which of the following types of fire is used to cover a gap between adjacent platoons?

1. Grazing fire
2. Plunging fire
3. Indirect fire
4. Direct fire

4-12. Which of the following information is NOT included in a platoon fire plan sketch?

1. On-call targets
2. Barriers
3. Location of listening posts
4. Location of the fire team leader

4-13. What officer establishes the order of priorities for organizing the platoon in the defense?

1. Company commander
2. Commanding officer
3. Operations officer
4. Supply officer

A.	Establish Security
B.	Prepare Alternate Positions
C.	Prepare Fighting Positions
D.	Assign Sectors of Fire

Figure 4A.—Priority task list.

IN ANSWERING QUESTIONS 4-14 THROUGH 4-17, REFER TO FIGURE 4A.

4-14. What task is performed as soon as you clear fields of fire?

1. A
2. B
3. C
4. D

4-15. What task should be performed first?

1. A
2. B
3. C
4. D

4-16. What task is performed as soon as you assign sectors of fire?

1. D
2. A
3. B
4. C

4-17. What task is normally done before establishing supplementary positions?

1. A
2. B
3. C
4. D

Learning Objective: Determine the best location for employing crew-served weapons and the role of each crew-served weapon in the defense.

A.	M60E Machine Gun
B.	.50 caliber Machine Gun
C.	M19 Grenade Launcher
D.	AT4 Antitank Weapon

Figure 4B.—Crew-served weapons.

IN ANSWERING QUESTIONS 4-18 THROUGH 4-21, REFER TO FIGURE 4B.

4-18. What crew-served weapon can be used for direct or indirect fire?

1. A
2. B
3. C
4. D

4-19. What crew-served weapon provides massive rapid fire for penetrating lightly armored vehicles?

1. A
2. B
3. C
4. D

4-20. What crew-served weapon serves as the basis for a FPF in the overall defensive fire plan?

1. A
2. B
3. C
4. D

4-21. What crew-served weapon is used primarily against armored personnel carriers?

1. A
2. B
3. C
4. D

22. A machine gun squad consists of a total of how many machine gun teams?

1. One
2. Two
3. Three
4. Four

23. A machine gun team consists of a total of how many personnel?

1. One
2. Two
3. Three
4. Four

24. Both machine guns of a squad are positioned a minimum of how many meters apart?

1. 20
2. 25
3. 30
4. 35

25. Normally, a sector of fire for a machine gun does not exceed a total of how many mils?

1. 800
2. 600
3. 500
4. 400

26. A machine gun cannot be assigned both a final protective line and a principle direction of fire.

1. True
2. False

27. Which of the following platoons normally have positioned assault squads?

1. Reserve platoons
2. Weapons platoons
3. Front-line platoons
4. Supplementary platoons

28. What means of communication does the weapons platoon commander use to communicate with assault squads?

1. Messenger
2. Radio
3. Landline
4. Hand signals

29. A single mortar is assigned what area of coverage for a final protective fire?

1. 25 by 50 meters
2. 100 by 100 meters
3. 30 by 30 meters
4. 50 by 50 meters

4-30. The final protective fire of a mortar should be directed at least how many meters from the FEBA?

1. 100
2. 60
3. 50
4. 40

4-31. A company is requesting a fire mission. What element of the battalion should they contact?

1. COC
2. FSC
3. FDC
4. BAS

4-32. What element within the battalion approves all fire missions?

1. COC
2. FDC
3. BAS
4. FSC

4-33. When you are firing a FPF without instructions, what is the rate of fire for the first 2 minutes for a mortar?

1. Ten rounds per minute
2. Nine rounds per minute
3. Eight rounds per minute
4. Five rounds per minute

Learning Objective: Explain the procedures for preparing a range card and the importance of the range card.

4-34. A total of how many range cards are prepared for each machine gun position?

1. One
2. Two
3. Three
4. Four

4-35. During defensive coordination planning, what weapon serves as the base from which to build the defense around?

1. The M203
2. The AT4
3. The mortar
4. The machine gun

4-36. In preparing a range card, you should use what type of format?

1. Standard format
2. Battalion's format
3. Universal format
4. Crew-served weapon format

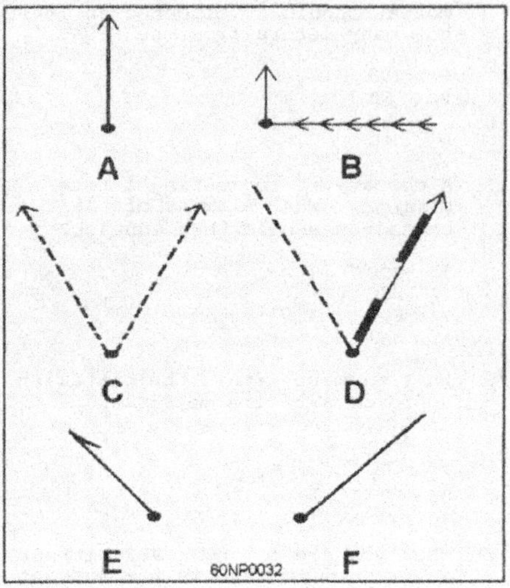

Figure 4C.—Range card symbols.

IN ANSWERING QUESTIONS 4-37 THROUGH 4-42, SELECT FROM FIGURE 4C THE PROPER RANGE CARD SYMBOL.

4-37. The symbol for a target line.

1. A
2. E
3. F
4. B

4-38. The symbol for magnetic orientation.

1. E
2. B
3. F
4. A

4-39. The symbol for grazing fire and dead space.

1. B
2. A
3. C
4. D

4-40. The symbol for sector limits.

1. D
2. C
3. B
4. E

4-41. The machine gun symbol.

1. A
2. E
3. C
4. B

4-42. The north arrow symbol.

1. F
2. A
3. E
4. B

4-43. There are two methods used on a range card to locate a gun positic positively. Which method uses a single-solid line drawn from a prominent terrain feature to the gun position with several arrow heads pointing in the direction of the gun position?

1. The location grid coordinates method
2. The magnetic orientation line method
3. The ground positioning system method
4. The UTM method

4-44. One degree is approximately equal to 17.7778 mils. An azimuth of 15 degrees is equal to how many mils.

1. 2,667 mils
2. 2,666 mils
3. 132 mils
4. 168 mils

4-45. Sector limits are drawn as broken lines ending in arrow heads. When a weapon is assigned an FPL, how many sector limit lines are drawn on a range card?

1. One
2. Two
3. Three
4. Four

4-46. Which of the following fire plans is complied and used primarily as the company fire plan?

1. Squad leaders' fire plan
2. Fire team leaders' fire plan
3. Weapons platoon commanders' fire plan
4. Platoon commanders' fire plan

Learning Objective: Identify the importance of counter ambush techniques. State the procedures for dismounting a vehicle and preparing a vehicle for a convoy.

4-47. Which of the following instructions provide guidance for convoy procedures?

1. COMSECONDNCB/COMTHIRDNCBINST 5200
2. COMSECONDNCB/COMTHIRDNCBINST 3122.1
3. COMSECONDNCB/COMTHIRDNCBINST 11200.1
4. COMSECONDNCB/COMTHIRDNCBINST 11200.11C

4-48. A common characteristic of an ambush is that an ambush will occur in how many phases?

1. One
2. Two
3. Three
4. Four

4-49. A total of how many sand bags is required to "harden" a 6 by 6 cargo truck?

1. 50 to 100
2. 70 to 100
3. 100 to 150
4. 100 to 200

4-50. When "hardening" the sides of a 6 by 6 cargo truck, you should stack sandbags how many layers high?

1. Five
2. Two
3. Three
4. Four

4-51. Which of the following personnel is normally designated as a vehicle commander in a convoy?

1. Fire team leader
2. Squad leader
3. Platoon commander
4. Assistant platoon commander

4-52. The shotgun rider for a vehicle in a convoy is normally armed with what type of weapon?

1. M60E3
2. AT4
3. .50 cal
4. M203

4-53. During maneuvers executed by the occupant squad of a vehicle, what maneuver is carried out by the shotgun rider?

1. Accompany the maneuvers executed by the occupant
2. Maneuver to protect the vehicle and the driver.
3. Remain seated in the vehicle
4. Maneuver to the rear of the vehicle

4-54. Which of the following personnel provide cover fire for the debarkation of occupants of a vehicle that is halted in the ambush killing area?

1. The squad leader
2. The fire team leader
3. The sentries
4. The platoon commander

4-55. When traveling with a squad of a convoy, the machine gun team is assigned what position in the vehicle?

1. Facing the front of the vehicle
2. Facing outboard on the left side of the vehicle
3. Facing outboard on the right side of the vehicle
4. Facing to the rear of the vehicle

4-56. What is the maximum number of people that should occupy the bed of a hardened 6 by 6 truck?

1. Ten
2. Eleven
3. Twelve
4. Thirteen

4-57. Which of the following convoy vehicle positions is NEVER occupied by the vehicle commander?

1. The center vehicle
2. The lead vehicle
3. The vehicle in front of the rear vehicle
4. The vehicle in back of the lead vehicle

4-58. Which of the following commanders must attend the convoy commander's briefing?

1. Platoon commanders
2. Weapons platoon commanders
3. Vehicle commanders
4. Commanding officer

4-59. Which of the following commanders designates a second in command for the convoy?

1. Vehicle commander
2. Platoon commander
3. Convoy commander
4. Commanding officer

Learning Objective: Determine the most effective course of action to take when a convoy encounters enemy activity.

4-60. A vehicle is fired upon in the killing zone. What action is taken by the driver?

1. Immediately halt the vehicle and disembark
2. Immediately attempt to drive through the killing zone
3. Immediately signal for help
4. Immediately inform the convoy commander

4-61. A "hardened" vehicle is forced to halt in the killing zone of an ambush. What action is taken by the occupants?

1. Remain with the vehicle at all times
2. Immediately dismount the vehicle
3. On the first perceptible slackening of enemy fire, dismount
4. Immediately inform the convoy commander

4-62. A "soft" vehicle is forced to halt in the killing zone of an ambush. What action is taken by the occupants?

1. Remain with the vehicle at all times
2. Immediately dismount the vehicle
3. On the first perceptible slackening of enemy fire, dismount
4. Immediately inform the convoy commander

4-63. What is the most effective counteraction from elements not in the killing zone of an ambush?

1. A flanking maneuver against the enemy
2. An immediate frontal assault against the enemy
3. An immediate withdrawal to a safe area
4. An immediate call for an air strike

4-64. What is the best counter ambush action for a armored vehicle, such as a tank?

1. Move into the killing zone
2. Move to the flanks of the enemy
3. Move to the rear of the killing zone
4. Move to the front of the killing zone

ASSIGNMENT 5

extbook Assignment: "Counter Ambush Techniques" and "CBR Decontamination," pages 5-6 through 6-17.

Learning Objective: Determine the most effective course of action to take when a convoy encounters enemy activity.

5-1. What course of action, if any, is taken when vehicles and cargo must be abandoned during a convoy?

1. Destroy all vehicles and cargo upon orders from the convoy commander
2. Move on with the convoy and leave all vehicles and cargo intact
3. Destroy all radios left behind and move on with the convoy
4. None

5-2. Which of the following guidelines is NOT effective in decreasing damage by mines in convoy operations?

1. Follow the tracks of the vehicle in front
2. Do not run over foreign objects, brush, or grass in the road
3. When possible drive on the shoulder of the road
4. Watch local national traffic and the reactions of people on foot

5-3. What course of action is taken when a convoy encounters sniper fire?

1. Aggressive action
2. Passive action
3. Destroy the sniper with short-range fire
4. Stop all vehicles and locate the sniper

5-4. Normally, small-arms fire is not effective against an air attack from an altitude equal to or greater than how many meters?

1. 100
2. 150
3. 200
4. 350

5-5. Personnel should stand air guard duty for what maximum time?

1. 15 minutes
2. 20 minutes
3. 30 minutes
4. 60 minutes

5-6. What factor is the key to effective small-arms fire against aircraft?

1. The volume of small-arms fire
2. The accuracy of small-arms fire
3. The position of the small-arms weapons in the convoy
4. Trying to shoot down an attacking airplane with small-arms weapons is fruitless

5-7. Which of the following is NOT a good tip for small-arms air defense?

1. Fire at the belly of an aircraft
2. Volume of fire is the key; everybody fires
3. Lie on your back if caught in the open so you can fire upon the aircraft
4. Mounted M-60s and M-2s should aim slightly above the nose of the aircraft for head-on targets

A.	Close column
B.	Open column
C.	Infiltration column
D.	Disperse column

Figure 5A.—Column formations.

IN ANSWERING QUESTIONS 5-8 THROUGH 5-13, REFER TO FIGURE 5A.

5-8. What column formation offers an advantage in that fewer vehicles are likely to be damaged from air-delivered rockets, cannons, or cluster bomb units?

1. A
2. B
3. C
4. D

5-9. What column formation prevents crowds from coming between the convoy and disrupting the movement of the convoy?

1. A
2. B
3. C
4. D

5-10. What column formation provides the best passive defense against hostile observation and attack?

1. A
2. B
3. C
4. D

5-11. What column formation makes control more difficult for the convoy commander to give orders?

1. A
2. B
3. C
4. D

5-12. What column formation provides the least active defense capability if individual or small groups of vehicles are attacked?

1. A
2. B
3. C
4. D

5-13. What column formation has a disadvantage in that, if attacked, its defense is less effective because small-arms fire is less concentrated?

1. A
2. B
3. C
4. D

5-14. During an air attack, the convoy commander chooses the option to stop the convoy. What is the disadvantage of this option?

1. It is harder for the pilot to see the convoy
2. It is harder to continue the march
3. The volume and density of organic weapons fire are less than when the convoy is dispersed
4. The stopped convoy makes a good target for enemy troops

5-15. During an air attack, the convoy commander chooses the option to disperse the convoy. What is the disadvantage of this option?

1. It is harder to communicate with the convoy
2. It is harder to call in a fire mission
3. The volume and density of the small-arms fire of the convoy are reduced
4. Makes it easier for the enemy pilot to detect the vehicles

---

Learning Objective: Explain the three types of immediate action drills and their importance.

5-16. What is the purpose of conducting immediate action drills?

1. They aid troops in reacting quickly and properly
2. They are a required PRCP skill
3. They are drills covering every contingency possible
4. They are drills that are stereotyped in nature

5-17. During a vehicle unloading drill, which of the following personnel provide a cover of fire while the occupants dismount?

1. The vehicle commander
2. The grenadier
3. The assistant driver
4. The sentries

5-18. Which of the following drills is undertaken when a unit has sighted guerrillas but the guerrillas have not sighted the unit?

1. Vehicle unloading drill
2. Freeze and hasty ambush drill
3. Immediate assault drill
4. Counter ambush drill

5-19. During a freeze and hasty ambush drill, you should move to what side of the trail if the enemy is traveling the same trail?

1. Left
2. Right
3. Same side as the originator spotting the enemy
4. Same side as the vehicle commander

20. During a freeze and hasty ambush drill, in which a unit member other than the unit leader is fired upon, what person initiates the ambush?

1. The unit leader
2. The individual being fired upon
3. The individual closer to the enemy
4. The vehicle commander

A.	Vehicle unloading drill
B.	Freeze and hasty ambush drill
C.	Immediate action drill
D.	Counter ambush drill

Figure 5B.—Immediate action drills.

N ANSWERING QUESTIONS 5-21 THROUGH 5-24, EFER TO FIGURE 5B.

21. A tactic used during an unexpected encounter at close quarters.

1. A
2. B
3. C
4. D

22. The drill undertaken when the unit has sighted guerrillas approaching but has not yet been seen by them.

1. A
2. B
3. C
4. D

23. The drill undertaken when the entire unit is in the kill zone of an ambush.

1. A
2. B
3. C
4. D

24. The drill undertaken when a vehicle is forced to halt in the killing zone of an ambush and the occupants must debark.

1. A
2. B
3. C
4. D

5-25. What person initiates an immediate assault drill?

1. The first member of the unit who sights the enemy
2. The unit leader
3. The convoy commander
4. The vehicle commander

5-26. When only the foremost elements of the unit are caught in the killing zone, what type of attack is used against the enemy?

1. Immediate frontal attack
2. Immediate flank attack
3. Immediate encircling attack
4. Immediate call for fire

Learning Objective: Explain the reasons for decontamination and the types and forms of contamination.

5-27. A battalion is required to have a minimum of how many six-man teams per company qualified to conduct decon operations?

1. One
2. Two
3. Three
4. Four

5-28. What are the three MAJOR forms that CBR contamination can be delivered?

1. Fine dust, dust pollinated, or powder
2. Mist, vapor, or rain
3. Dust pollinated, mist, or vapor
4. Solids, liquids, or gases

5-29. Generally, when a transfer or spread hazard exists, a vapor hazard could also exist.

1. True
2. False

5-30. What MINIMUM protection is required if you are operating a vehicle that has absorbed a nerve agent?

1. Full MOPP gear
2. A protective mask only
3. A protective mask and gloves
4. Gloves only

5-31. For decon purposes, radiation can be thought of as existing in what state of matter?

1. Liquid
2. Solid
3. Gases
4. Vapor

5-32. Which of the following ranges does the AN/PDR-27 instrument provide for radioactivity detection?

1. Low range only
2. Medium range only
3. High range only
4. Low through high ranges

5-33. Which of the following ranges does the AN/PDR-43 instrument provide for detection of radioactivity?

1. Low range
2. Medium range
3. High range
4. Low through high ranges

5-34. Using the "7-10 rule of thumb," if 3 hours after a burst your radiation reading is 100 Centigray (cGy) (rad) per hour, then how many hours after the burst can you expect a reading of about 10 (cGy)?

1. 14
2. 21
3. 30
4. 70

5-35. Negligible risk levels for biological and chemical contamination are contamination that will cause mild incapacitation among no more than what percent of unprotected troops operating for 12 continuous hours within 1 meter of a contaminated surface?

1. 5%
2. 10%
3. 15%
4. 20%

5-36. Which of the following measurements for radiological contamination is considered a negligible risk level?

1. .10 cGy or less
2. .50 cgy or less
3. .40 cGy or less
4. .33 cGy or less

5-37. Which of the following agents should you decontaminate first if your unit is contaminated by a mixture of agents?

1. Chemical agent
2. Biological agent
3. Radiological agent
4. Biological and radiological agents

5-33. The decon methods for biological agents are also effective for neutralizing or removing chemical and radiological contamination.

1. True
2. False

5-39. What are appropriate decon measure based on?

1. The first hazard identified
2. The first chemical hazard identified
3. The first radiological hazard identified
4. After thoroughly identifying all agent hazards

5-40. Within 30 days after the initial opening of the original protective package, the chemical protective overgarment (CPO) is good for a total of how many hours in an UNCONTAMINATED environment?

1. 120 hours
2. 130 hours
3. 140 hours
4. 150 hours

5-41. What happens to an overgarment after it comes in contact with Pols?

1. Overgarment must be replaced
2. The time the overgarment is good for is increased
3. The time the overgarment is good for is decreased
4. The overgarment must be cleanec immediately

5-42. When wearing a chemical protective overgarment for a cumulative wear time of 138 hours and the environment becomes CONTAMINATED, you may wear the suit for how many additional hours?

1. 8
2. 2
3. 6
4. 4

5-43. Chemical protective overgarment provides protection in a CONTAMINATED environment for what maximum period of time?

1. 8 hours
2. 2 hours
3. 6 hours
4. 4 hours

5-44. The battle dress overgarment (BDO), currently not in the battalion's TOA, provides protection for how many hours after becoming contaminated with liquid chemical agents?

1. 12
2. 14
3. 20
4. 24

Learning Objective: Identify the four principles of DECON.

A. Decontaminate as soon as possible
B. Decontaminate only what is necessary
C. Decontaminate as far forward as possible
D. Decontaminate by priority

Figure 5C.—Principles of DECON.

IN ANSWERING QUESTIONS 5-45 THROUGH 5-49, REFER TO FIGURE 5C.

5-45. Which principle allows decon to begin earlier and limits the spread of contamination to other areas?

1. A
2. B
3. C
4. D

5-46. Which principle consists of cleaning important items of equipment first and the least important items last?

1. A
2. B
3. C
4. D

5-47. Which principle emphasizes not to waste precious resources by decontaminating everything?

1. A
2. B
3. C
4. D

5-48. Of the four, which principle is considered the most important?

1. A
2. B
3. C
4. D

5-49. Which principle must be considered first before you consider any other principle?

1. A
2. B
3. C
4. D

Learning Objective: Differentiate the three levels of DECON.

5-50. Which level of DECON aims at sustaining operations, reducing the contact hazard, and limiting the spread of contamination?

1. Immediate DECON
2. Operational DECON
3. Thorough DECON
4. Detailed equipment DECON

5-51. What level of DECON is designed to reduce contamination on personnel, equipment/material, and/or working areas to the lowest possible level (negligible risk)?

1. Operational DECON
2. Thorough DECON
3. Immediate DECON
4. Personnel wipe down

5-52. Which level of DECON includes skin DECON?

1. Thorough DECON
2. Operational DECON
3. Operator's spray down
4. Immediate DECON

5-53. Which level of DECON includes detailed troop DECON (DTD)?

1. Thorough DECON
2. Operational DECON
3. Immediate DECON
4. Operator's spray down

5-54. What level of DECON includes the DECON technique, vehicle wash down?

1. Immediate DECON
2. Operational DECON
3. Thorough DECON
4. Operator's spray down

Learning Objective: Explain immediate DECON operations.

5-55. What person initiates immediate DECON?

1. Commanding officer
2. Platoon commander
3. Company commander
4. The individual

5-56. What is the PRIMARY purpose of the M291 DECON kit?

1. To decon exposed skin
2. To decon equipment
3. To decon weapons
4. To decon MOPP gear

5-57. What is the PRIMARY purpose of the M280DKIE DECON kit?

1. To decon vehicles
2. To decon exposed skin
3. To decon individual equipment
4. To decon personnel

5-58. DECON kits M291/M258A1 are used to remove what type of contamination?

1. Chemical contamination
2. Biological contamination
3. Chemical and biological contamination
4. Radiological contamination

5-59. Which of the following is a method for removing radiological contamination?

1. Apply soap and water
2. The M2581A DECON kit
3. The M291 DECON kit
4. The M280DKIE DECON kit

5-60. You should initiate the skin decon technique within how many minutes of becoming contaminated?

1. 1
2. 5
3. 10
4. 15

5-61. At what temperature does the solutions in the M258A1 decon kit become unstable?

1. 100°F
2. 110°F
3. 115°F
4. 120°F

5-62. The personal wipe down technique is most effective when done within how many minutes of being contaminated?

1. 5
2. 10
3. 15
4. 20

5-63. What action should you take if a chemical agent is globbed on your overgarment?

1. Use your M2581 DECON kit
2. Use a stick to scrape off the chemical agent
3. Use your M291 DECON kit
4. Use your M280DKIE DECON kit

5-64. You should not make an attempt to decontaminate chemical agents on your overgarment.

1. True
2. False

5-65. Operator's spray down is most effective when done within how many minutes after personal wipe down?

1. 10
2. 15
3. 20
4. 25

5-66. After scrubbing DS2 onto a CHEMICAL contaminated surface, how long should you wait before washing it off?

1. 30 minutes
2. 20 minutes
3. 15 minutes
4. 5 minutes

5-67. Do not mix DS2 and STB because mixing the two will cause a fire.

1. True
2. False

ASSIGNMENT 6

Textbook Assignment: "CBR Decontamination," pages 6-17 through 6-50.

Learning Objective: Explain the preparation, execution, and site clearance phases of operational decon.

6-1. What is the primary objective of operational decon?

1. To reduce the level of contamination to regenerate needed combat power
2. To eliminate contamination
3. To decontaminate the TOA
4. To decontaminate personnel

A.	M-17
B.	ICAD
C.	M13
D.	M11
E.	CAM

Figure 6A.—Decon assets.

IN ANSWERING QUESTIONS 6-2 THROUGH 6-6, REFER TO FIGURE 6A.

6-2. What DECON asset is used as a monitor for chemical agents?

1. A
2. B
3. C
4. E

6-3. What DECON asset is used for detecting individual chemical agents?

1. A
2. B
3. D
4. E

6-4. What DECON asset contains a personnel shower assembly?

1. A
2. B
3. C
4. D

6-5. What DECON asset is a brush apparatus for scrubbing with DS2?

1. A
2. B
3. C
4. D

6-6. What DECON asset is a fire extinguisher-like device for spraying DS2?

1. B
2. C
3. D
4. E

6-7. During operational decon, what two operations are accomplished in conjunction?

1. Vehicle wash down and personnel wipe down
2. Operator's spray down and personnel wipe down
3. Skin decontamination and vehicle wash down
4. MOPP gear exchange and vehicle wash down

6-8. When planning a water source for an operational decon site, you should consider how many gallons of water per vehicle?

1. 50 to 100
2. 100 to 150
3. 150 to 200
4. 200 to 250

6-9. A M12A1 PDDA can be used to transport a total of how many gallons of water to a decon site?

1. 200
2. 250
3. 300
4. 450

6-10 The M17 LDS collapsible bladder must be filled up on site. It can hold a total of how many gallons of water?

1. 500
2. 1,000
3. 1,500
4. 2,000

6-11. The MOPP gear exchange site is located how many meters upwind from the vehicle wash down site?

1. 50
2. 75
3. 100
4. 150

6-12. What are the two operational decon techniques?

1. Vehicle wash down and MOPP gear exchange
2. Skin decontamination and vehicle wash down
3. Operator's spray down and personnel wipe down
4. Vehicle wash down and personnel wipe down

6-13. Personnel in wheeled vehicles should take what action before vehicle wash down?

1. Assist in security operations
2. Dismount the vehicle to avoid getting wet
3. Stay in the vehicle to ensure all decontaminants are removed
4. Assist the decon team

6-14. For planning purposes, the vehicle wash down site will process how many vehicles every 2 to 3 minutes?

1. One
2. Two
3. Three
4. Four

6-15. During operational decon how much time is planned for MOPP gear exchange?

1. 10 minutes
2. 15 minutes
3. 20 minutes
4. 30 minutes

6-16. What is done before vehicle wash down to increase decon effectiveness of vehicles?

1. MOPP gear exchange
2. Skin decontamination
3. Personnel wipe down
4. Operator's spray down

6-17. What is the range of the water pressure that is needed to remove most gross contamination?

1. 30 to 60 psi
2. 40 to 100 psi
3. 50 to 120 psi
4. 60 to 120 psi

6-18. If the mission permits, vehicle wash down is most effective during what time period?

1. 1 to 6 hours after contamination
2. 7 to 8 hours after contamination
3. 9 to 12 hours after contamination
4. 14 to 16 hours after contamination

Learning Objective: Explain the steps for performing MOPP gear exchange.

6-19. MOPP gear exchange is performed by units that are what size?

1. A company or a platoon
2. A squad or a crew
3. A fire team
4. Individually

6-20. Normally, a total of how many steps is used during MOPP gear exchange?

1. Five
2. Six
3. Seven
4. Eight

A.	Step 1, decon gear
B.	Step 2, decon mask and gloves
C.	Step 5, don new overgarment
D.	Step 4, remove overboots and gloves
E.	Step 6, don overboots and gloves

Figure 6B.—Steps for MOPP gear exchange.

IN ANSWERING QUESTIONS 6-21 THROUGH 6-25, REFER TO FIGURE 6B.

6-21. What step involves a buddy opening a smock package without touching the inside?

1. A
2. B
3. C
4. D

22. What step involves rubbing a dry mix of STB into personnel gear, such as your weapon and helmet?

1. A
2. B
3. C
4. D

23. What step involves the member to remove the inner and outer gloves without touching the outside of the package?

1. B
2. C
3. D
4. E

24. What step involves a buddy to decon a member's mask?

1. A
2. B
3. D
4. E

25. What step involves the member to untie or cut the laces on their boots?

1. A
2. B
3. C
4. D

26. Techniques used for MOPP gear exchange enables troops to unmask.

1. True
2. False

Learning Objective: Identify the four phases of thorough decon.

---

27. A typical vehicle requires how many gallons of water during detailed equipment decon?

1. 300
2. 400
3. 500
4. 600

28. What is the primary purpose of the predecon staging area?

1. To prepare the contaminated unit for DED and DTD operations
2. To enable the contaminated unit to unmask
3. To prepare the contaminated unit for MOPP gear exchange
4. To prepare the contaminated unit for DTD

6-29. When using a chemical agent monitor (CAM), you should ensure there is sufficient distance between vehicles for what reason?

1. For safety reasons
2. Allows the crew room to dismount
3. Vapor drift can give a false reading
4. Enables you to work around the vehicles

6-30. In preparation for detailed equipment decon (DED), crews should concentrate on removing mud and debris from what area of a vehicle?

1. The top of the vehicle
2. The sides of vehicle
3. The engine compartment
4. The undercarriage

6-31. Before DED operations, what should you do with materials on a vehicle that can absorb liquid contamination?

1. Remove and place at a collection point
2. Secure firmly to the vehicle
3. Remove and discard
4. Remove and bury

6-32. What group of individuals are sent thorough the DTD first?

1. Squad leaders
2. Platoon commanders
3. Company commanders
4. Assistant vehicle drivers

6-33. What action is considered major in conducting detailed troop decon (DED)?

1. Removing contamination from vehicles and gear
2. Removing contaminated MOPP gear, including the protective mask
3. Removing contaminated MOPP gear only
4. Removing contaminated protective mask only

6-34. What organization is responsible for setting up, operating, and closing the DTD in a thorough decon site?

1. The decon team
2. The contaminated unit
3. The contaminated platoon
4. The contaminated squad

6-35. A total of how many stations are included in a detailed troop decon (DTD)?

1. 6
2. 7
3. 8
4. 9

6-36. Station 1 of DTD requires how many pounds of STB to be mixed with 20 gallons of hot water?

1. 50
2. 100
3. 150
4. 200

6-37. A poncho divided in half and placed at DTD station 1 serves what purpose?

1. A place for the removal of contaminated gear
2. A place for the removal of uncontaminated gear
3. A place for the removal of decontaminated boots
4. As a contamination control line (CCL).

6-38. DTD station 1 is located a minimum of 3 meters from all other stations for what reason?

1. To avoid congestion
2. To ensure a true reading on detection equipment
3. To ensure troops have enough space to perform MOPP gear exchange
4. To ensure troops have enough space to perform immediate decon

6-39. What decon activity is conducted at DTD station 2?

1. Overboot and hood decon
2. MOPP gear decon
3. Mask decon
4. Mask and gloves decon

6-40. Ten people have been processed through the shuffle pit at DTD station 2. What should you do to the mix in the shuffle pit?

1. Bury the STB mix
2. Replace the STB mix
3. Add more STB to the mix
4. Remove half of the STB mix

6-41. Which of the following station is notified if a punctured boot is found at DTD station 2?

1. Station 5
2. Station 2
3. Station 3
4. Station 4

6-42. How many decon team members are needed to decontaminate the overboots and gloves at DTD station 4?

1. One
2. Two
3. Three
4. Four

6-43. What is the purpose of the liquid contamination control line (LCL)?

1. Holding location for contaminated troops
2. Holding location for uncontaminated troops
3. Drop-off point for contaminated gear
4. To separate the "dirty" from the "clean" areas

6-44. What is one of the primary functions of DTD station 5?

1. To decon hood and gloves
2. To decon boots
3. To decon the protective mask
4. To treat any personnel suffering from chemical agents symptoms

6-45. What is the primary function of DTD station 6?

1. To decon MOPP gear
2. To decon boots
3. Mask removal
4. To decon the protective mask

6-46. After ten masks have been dipped in a sanitizing solution at station 6, you should take what action?

1. Discard the solution into a sump
2. Add some more solution
3. Add some calcium hypochlorite
4. Add some STB to the solution

6-47. What are the three contamination control lines that are established when setting up for detailed troop decon?

1. CCL, LCL, and VCL
2. DTD, DED, and ID
3. LCD, CLL, and VCL
4. LCL, DCL, and VCL

6-48. Which of the companies is normally responsible for the setup of the DED site?

1. Alpha
2. Bravo
3. Charlie
4. Hotel

6-49. For radiological contamination, which of the following station numbers is not used during DED operations?

1. 1
2. 2
3. 3
4. 4

6-50. At what temperature should hot, soapy water be heated?

1. 80°F to 100°F
2. 100°F to 120°F
3. 120°F to 140°F
4. 140°F to 160°F

6-51. The runoff from station 1 of detailed equipment decon (DED) is contaminated and must be treated as hazardous.

1. True
2. False

6-52. In order to limit the work load of each member of the scrubbing team at DED station 2, a vehicle is divided into how many parts for scrubbing with DS2?

1. One
2. Two
3. Three
4. Four

6-53. After a vehicle is applied with DS2, how many minutes should the vehicle remain at DED station 3?

1. 15
2. 20
3. 30
4. 60

6-54. When, if ever, should you mix HTH or STB with DS2?

1. When heavy contamination is removed
2. When radiological contamination is removed
3. When mud is removed
4. NEVER

6-55. Once interior decon is completed at DED station 3, the driver of the vehicle proceeds to what station?

1. DTD, station 1
2. DED, station 2
3. DTD, station 3
4. DED, station 4

6-56. What person moves a vehicle from DED station 3 to DED station 4?

1. A decon team member
2. A crew member
3. A assistant driver
4. The POIC

6-57. To turn 5 gallons of water into a 10 percent DECON solution, you must add how many pounds of STB or HTH?

1. 2.5 lb
2. 3.5 lb
3. 4.5 lb
4. 5.5 lb

6-58. What is the primary task of DED station 5?

1. To apply DS2 to a vehicle
2. To apply STB to a vehicle
3. To remove HTH from a vehicle
4. To remove DS2 from a vehicle

6-59. Significant chemical contamination is found on a vehicle at DED station 5. This vehicle should be recycled to what DED station?

1. 1
2. 2
3. 3
4. 4

Learning Objective: Identify the different detailed equipment decon site configurations.

6-60. The optimum setup of a DED for a M12A1 PDDA-equipped decon platoon can process what maximum number of vehicles per hour?

1. Five
2. Six
3. Seven
4. Eight

Made in the USA
Monee, IL
13 September 2026

62976963R00105